Lengsfeld, Wolff-Fabris, Krämer, Lacalle, Altstädt

Composite Technology

Hauke Lengsfeld
Felipe Wolff-Fabris
Johannes Krämer
Javier Lacalle
Volker Altstädt

Composite Technology

Prepregs and Monolithic Part Fabrication Technologies

HANSER

Hanser Publishers, Munich Hanser Publications, Cincinnati

The Authors:

Hauke Lengsfeld, Hexcel Composites GmbH, Stade, Germany

Felipe Wolff-Fabris, SKZ - Das Kunststoff-Zentrum/Europäisches Zentrum für Dispersionstechnologie (EZD), Selb, Germany

Johannes Krämer, BASF SE, Ludwigshafen, Germany

Javier Lacalle, MTorres Deutschland GmbH, Stade, Germany

Volker Altstädt, Universität Bayreuth, Lehrstuhl für Polymere Werkstoffe, Bayreuth, Germany

Distributed in the Americas by:
Hanser Publications
6915 Valley Avenue, Cincinnati, Ohio 45244-3029, USA
Fax: (513) 527-8801
Phone: (513) 527-8977
www.hanserpublications.com

Distributed in all other countries by:
Carl Hanser Verlag
Postfach 86 04 20, 81631 München, Germany
Fax: +49 (89) 98 48 09
www.hanser-fachbuch.de

Cataloging-in-Publication Data is on file with the Library of Congress

Editor: Mark Smith
Production Management: Jörg Strohbach
Coverconcept: Marc Müller-Bremer, www.rebranding.de, München
Coverdesign: Stephan Rönigk
Typesetting: Kösel Media, Krugzell
Printed and bound by Hubert & Co GmbH, Göttingen, Germany

ISBN: 978-1-56990-599-9
E-Book ISBN: 978-1-56990-600-2

Preface

Prepregs have famous predecessors, such as the plywood found in 3400 BC in ancient Mesopotamia. The Mesopotamians realized that glueing wood at different angles provided natural wood with better properties. Also papyrus, an early type of paper made from the stem of the papyrus plant, has a visible, fibrous, unidirectional reinforcement and was used in ancient Egypt (2500 BC) as a writing surface.

Although long declared obsolete, prepregs have gained increasing popularity in all segments of the composite industry because of their versatility, high fiber volume content, and large variety of fiber-matrix combinations.

Beginning in the early 1970s until today, structural carbon fiber/plastic components have been manufactured using prepreg technology. The authors therefore felt that it was time to provide a current overview of trends and perspectives relating to this technology.

Thermosetting prepregs are manufactured by impregnating reinforcing fibers or textiles with custom formulated, pre-catalyzed resin systems. The machines and technology used for this impregnation facilitate close contact between fiber and matrix and a defined surface tack of the manufactured tapes. These pregregs are used to manufacture composite components faster and with less performance loss than can be achieved with similar wet-impregnation technologies. Covered with pliable carrier paper, the prepregs are easy to handle and remain flexible for a certain period of time at room temperature.

Globally, the market for prepregs has grown by 10% since 2012; in fact, it doubled in size in the period between 2002 and 2007 (see Lucintel's Global Prepreg Market 2008–2013: Trends, Forecast and Opportunity Analysis). Although the operational costs of manufacturing prepregs are often higher than those of traditional materials, their use may provide significant cost advantages when considering their weight-specific properties.

This book covers the important advances made in research and development, both in academic and industrial laboratories. It identifies the fundamental relationships between material structure, processing, and material properties. Looking ahead, the book introduces significant developments of modern prepreg technology.

After a general introduction into the world of fiber reinforced materials (Chapter 1), the fundamental prepreg components are introduced in Chapter 2. The following two chapters describe how prepreg materials are produced and subsequently used to manufacture preforms (Chapters 3 and 4). Chapters 5 and 6 explain how these preforms are cured into final components using autoclaves, ovens, or Quickstep technology. They also illustrate the technologies of the respective curing tooling. The cured components are subsequently tested and then typically assembled to form complex structures. Chapters 7 and 8 highlight the specific characteristics to be considered when using prepreg technologies.

The book aims at providing a comprehensive, application-oriented overview to the expert that not only clearly introduces the current state of the technology, but also illustrates its historic development.

Bayreuth, July 2015 *Volker Altstädt*

Acknowledgements

We would like to express our gratitude to all who facilitated the publication of this book.

Thanks to Carl Hanser Publishers who supported the publication of this book, in particular Ms. Ulrike Wittmann for the design and formatting of the chapters.

Thanks to Hans Marquardt and Dr. Wolfgang Nemetz, Airbus Operations GmbH, Dr. Hilmar Apmann, Premium Aerotec GmbH, and Michaela Kersch and Thomas Neumeyer, University of Bayreuth for their technical suggestions, discussions, and illustrations as well as to Mike Turner for his input on the history of prepreg development in the UK.

Particular thanks go to MTorres and Hexcel Composites, the companies who provided the major part of the illustrations.

Thank you also to our families for their patience and support.

Contents

Abbreviations and Important Terms

Activation	Heating of a prepreg to increase tack
AFP	Automated fiber placement
ALV	Average lay-up velocity
ATL	Automated tape laying
BMC	Bulk molding compound
BMI	Bismaleimide
BOX	Benzoxazine
CAD	Computer aided design
CAM	Computer aided manufacturing
CFM	Continuous filament mat
CFRP	Carbon fiber reinforced plastic
CNC	Computerized numerical control
CPT	Cured ply thickness
CSM	Chopped strand mat
CTE	Coefficient of thermal expansion
DD	Double diaphragm
DDS	Diaminodiphenyl sulfone
DFP	Dry fiber placement
DGEBA	Diglycidyl ether of bisphenol A
DICY	Dicyandiamide
DSC	Differental scanning calorimetry
EP	Epoxy
FAW	Fiber aerial weight
FRC	Fiber reinforced composite
FRP	Fiber reinforced plastics
FVC	Fiber volume content
FW	Filament winding
GFRP	Glass fiber reinforced plastic
HLU	Hand lay-up
HM	High modulus

HMI	Human machine interface
HS	High strength
HT fiber	High tensile fiber
IM	Intermediate modulus
IM fiber	Intermediate modulus fiber
Lay-up	Sequential deposition of flat semi-finished products in individual layers
NCF	Non-crimped fabric
ME roving	Multi-end roving
NC	Numerical control
NGT	Nickel-galvano tooling
OoA	Out of autoclave
Overlap/Gap	Overlap/gap between two parallel tapes
PA	Polyamide
PAW	Prepreg aerial weight
PE	Polyethylene
PEEK	Polyether ether ketone
PEI	Polyether imide
PES	Polyether sulfone
PLC	Programmable logic controller
Ply book	Describes the sequence of individual plies in a part
PP	Polypropylene
Prepreg	**Pre**im**preg**nated fibers
Prepreg tow	Narrow prepreg tape
RC	Resin content
RF	Resin flow
RTM	Resin transfer molding
SD	Single diaphragm
SE roving	Single-end roving
Slit tape	Unidirectional, narrow tape (prepreg) that was cut from a wider fiber tape
SMC	Sheet molding compound
SQRTM	Same qualified resin transfer molding
Stacking	Set fiber orientation and sequence
Staggering	Relative distance between tapes in two layers/plies with the same orientation
Staging	Degree of cross linking (α)
Steering	Draping in the plane
TCP	Tool center point
TGDDM	Alternative abbreviation for TGMDA
TGMDA	*N*,*N*,*N*',*N*'-Tetraglycidyl-4,4'-diaminodiphenylmethane
Tow	Bundle of carbon fiber filaments
UD	Unidirectional

UP	Unsaturated polyester resins
VC	Volatile content
WPU	Water pickup
WR	Woven roving

1 Introduction

Felipe Wolff-Fabris

1.1 Materials

The discovery and development of new materials has always had societal, economical, and military impact on mankind. Clay figurines were fired as early as in the Mesolithic Age, approx. 40,000 to 8500 BC. Metals also look back on a long history. The Copper Age encompassed the period from the Neolithic Age to the time when tools and ornaments were made (5000 to 3000 BC). At this time, agriculture took hold and mankind made its first tentative steps in the arts. The periods after the Copper Age are also identified by metals: the Bronze Age, named after an alloy that was discovered by chance, spanned from 3000 to 1000 BC, and the Iron Age from 1000 BC to modern times. The manufacturing and processing of materials allowed for the advancement of the intrinsic imagination of mankind. It was possible to transfer ideas into the physical, tangible world.

The last 150 years have seen unparalleled population increase and breathtaking technological development never seen before in human history. The industrial revolution, i.e., the large-scale adoption of manufacturing machines, was a major contributor to this development. Research and development resulted in new materials that now are indispensable parts of our everyday life. An example is the use of novel aluminum alloys that allow for the manufacture of lighter, larger, and better-performing airplanes. Also, the extremely high thermal resistance of technical ceramics (up to 3000 °C) facilitated space exploration.

Natural plastic materials, such as natural rubber and amber, have been known for centuries. In 1905, as one of the most important milestones in the history of plastic materials, Leo Hendrik Baekeland developed a process that used pressure and temperature to initiate a chemical reaction between two liquids (phenol and formaldehyde), creating a solid material. The resulting phenolic resin, commercially available a few years later under the trade name Bakelite, was used for telephone housings. However, numerous synthetic materials in this latest class of materials were developed only during the last decades.

Compared to metals and ceramics, the shaping of plastic materials requires significantly less energy. Plastics also exhibit considerably lower densities. Because of these advantages, plastic materials have substituted metals and ceramics in numerous applications, and they now facilitate the manufacturing of completely novel products. Here again, mankind has a new tool at its disposal to transfer our most ingenious ideas into the physical world. From wireless communication to driverless cars, there are no limits to our creativity. Future historians may look back and state: The Plastic Age began around 2000 AD.

Table 1.1 presents a general comparison of the advantages and disadvantages of metals, ceramics, and plastic materials. The data on global production emphasize the immense role of plastic materials.

Table 1.1 General Properties of Metals, Ceramics, and Plastics

Material	Pros	Cons	Estimated world production in 2012 [1–5]
Metals	▪ High toughness ▪ High stiffness and strength	▪ High density	1615 Mt/217 million m^3[a]
Ceramics	▪ Excellent temperature resistance ▪ High stiffness	▪ Low toughness	200 Mt/76 million m^3[b]
Plastics	▪ Low density ▪ Easy to process	▪ Poor temperature resistance	288 Mt/288 million m^3[c]

[a]: Crude steel, aluminum, copper, and titanium.
[b]: Glass, porcelain, and technical ceramics
[c]: Assumption: $\rho = 1000\ \text{kg/m}^3$

On the molecular level, plastic materials consist of repeating chemical building blocks; they are therefore also referred to as polymers. The Greek *polymer* means "many parts" (*poly*: many, *meros*: parts). Depending on their chemical structure, polymers are categorized as thermoplastics or thermosets.

Thermoplastic materials consist of very long molecular chains, formed by repeating units as mentioned earlier. Their solid-like behavior is caused by the entanglement of individual molecular chains. Examples for this class of materials include polypropylene (PP), polystyrene (PS), polyethylene (PE), and polyamide (PA), among others. Phenolic resins, as well as polyester and epoxy resin, are examples of thermosetting materials. Thermosets also consist of molecular chains; however, here these chains are cross-linked. In cross-linked molecular chains, chemical bonds not only hold together the chains, they are also formed between chains. A comprehensive discussion of thermoplastics and thermosets can be found in the literature [6].

Combination of Materials

The combination of existing materials offers the possibility to create new and improved technical materials. It is also possible to impart novel property profiles by using mineral or metallic additives. In a broader sense, all materials consisting of two or more substances can be defined as composite materials.

Fibrous materials, i.e., materials with a high aspect ratio (length/diameter), offer the potential to significantly enhance the mechanical properties of the continuous phase to be reinforced (the so-called matrix). The fibers act like a frame, bearing external loads. The continuous matrix distributes these loads and holds the fibers in place. The combination of matrix and fibers results in new property profiles that cannot be achieved by the individual components. Or, as Aristotle is quoted to have said more than 2000 years ago: *The whole is greater than the sum of its parts.* The same is true for the active principle of a fiber reinforced material.

The reinforcing principle of fiber reinforced materials has long been observed in nature. Wood, one of the oldest construction materials, exhibits excellent specific properties and to this day is irreplaceable in many applications. The main constituents of wood are cellulose, lignin, and polyose. Cellulose fibrils within the cellular structure act as reinforcing agents; they are held in place by a lignin matrix. Here, polyose serves as a bonding agent (coupling agent) between cellulose and lignin.

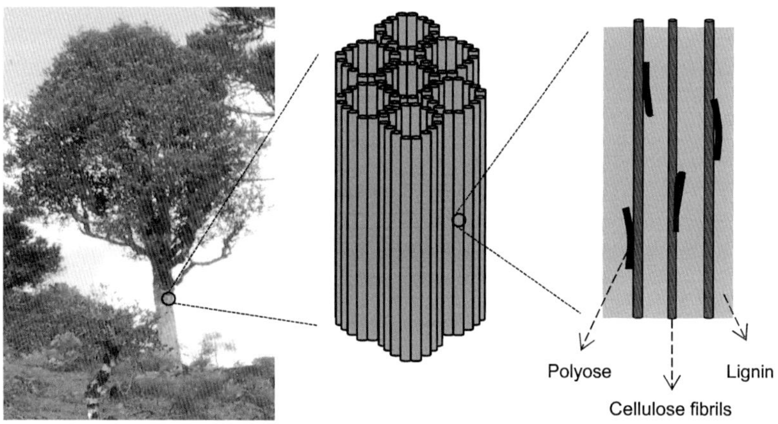

Figure 1.1 Schematic structure of wood

In order to combine the advantages of plastics and ceramics, polymeric materials are primarily reinforced using ceramic fibers (Table 1.1). They include glass and carbon fibers (see Chapter 2). This fiber reinforcement provides not only improved mechanical properties, it also increases the heat resistance of plastic materials and thus makes plastics suitable for expanded areas of application.

Both the length and orientation of the individual fibers determine the final properties of a fiber reinforced material. Figure 1.2 schematically shows the effect of various types of fiber reinforcement on the strength of the respective plastic material.

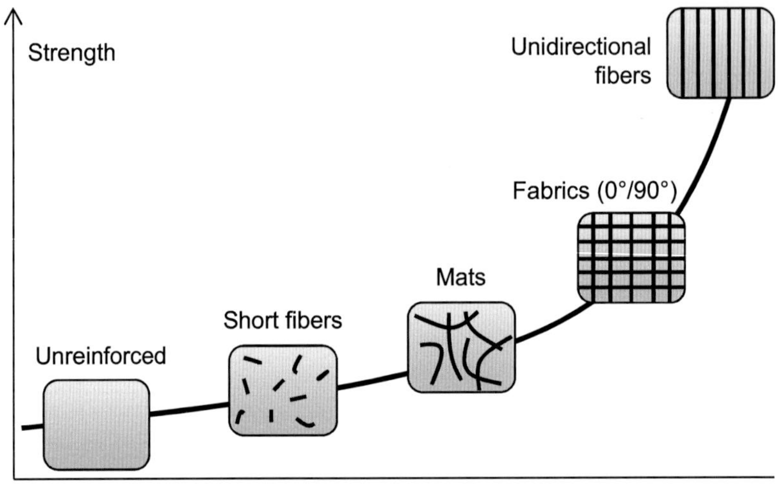

Figure 1.2 Effect of reinforcement type on strength of plastic materials

The market for fiber reinforced materials is continuously growing. It includes automotive parts made from short-fiber reinforced thermoplastics as well as high-performance components made from continuous fiber reinforced thermosets for the aviation industry.

■ 1.2 Continuous Fiber Reinforced Plastics

1.2.1 Property Profile

Fiber type, content, length, and orientation determine the degree of reinforcement in a plastic material. Generally, we distinguish between short- (< 1 mm), long- (from 1 to 50 mm), and continuous fiber reinforced plastic materials. The latter are multi-component systems consisting of a continuous, high-tensile fiber (e.g., glass or carbon fiber) and a tough polymeric matrix [7].

In order to maintain the reinforcing effect of the fiber in the plastic material, the following requirements have to be met: strength and Young's modulus of the fiber material must be higher than those of the matrix polymer. However, the matrix material has to exhibit higher elongation at break than the fiber.

Within the composite, the fibers absorb external loads, which requires high stiffness and strength. For lightweight construction applications, the fibers should also exhibit low densities. Chapter 2 will discuss the relevant reinforcing fibers and their properties in more detail.

The matrix has to fulfill a number of important tasks: it allows molding of the material, binds and positions the reinforcing fibers, and transfers loads from fiber to fiber. Chapter 2 will also introduce the relevant thermosets that can be used as matrix materials in fiber reinforced plastics.

However, in order to achieve the desired mechanical properties in the combination of matrix and fibers, the above requirements are not sufficient. Weak adhesive bonds between both components will negate the advantage of either fiber strength or load transfer by the matrix. Interfacial fiber-matrix adhesion is crucial for the quality and end-use characteristics of fiber reinforced plastic components.

Given sufficient adhesion between fiber and matrix, the mechanical properties of fiber reinforced materials are mainly determined by fiber orientation. In general, fiber reinforced materials exhibit anisotropic behavior, i.e., their mechanical properties change with the orientation of the load. They exhibit their highest possible strength and stiffness under tensile loads parallel to fiber orientation. However, tensile loads perpendicular to fiber orientation result in low matrix reinforcement (stiffness) and strength may even decrease. The dependence of the mechanical behavior of fiber reinforced plastics on fiber orientation is well documented in the literature [8].

The anisotropic behavior of fiber reinforced plastics represents a major difference compared to other common materials, such as aluminum and un-reinforced plastics, and has to be carefully considered when designing with fiber reinforced plastics. In order to properly dimension a fiber reinforced component and to adjust the fiber orientation, the prevailing types of loads and load directions encountered during component use have to be known (or at least be estimated). This ensures that fiber reinforced components are designed appropriately to withstand the loads encountered during use, and that the possibilities and advantages of this material class are fully utilized. Chapter 8 will describe a variety of design concepts for the manufacture of high-performance fiber reinforced components in more detail.

Material-specific properties of fiber reinforced materials as well as of other commonly used lightweight construction materials are shown in Table 1.2. The very high specific properties, i.e., those with respect to material densities, exemplify the enormous potential of fiber reinforced materials for lightweight applications.

Table 1.2 Typical Specific Young's Modulus (ρ) and Strength (σ/ρ) of Selected Lightweight Construction Materials

Material	Specific Young's modulus (GPa/g cm^{-3})	Specific strength (MPa/g cm^{-3})
Aluminum alloy	20–30	150–200
Titanium alloy	20–25	300–400
Glass fiber reinforced epoxy (unidirectional)	20–25	600–800
Glass fiber reinforced epoxy (quasi-isotropic)	10–12	200–250
Carbon fiber reinforced epoxy (unidirectional)	70–100	1200–1800
Carbon fiber reinforced epoxy (quasi-isotropic)	30–50	250–300

The correct determination of the mechanical properties of fiber reinforced materials requires highly qualified and experienced personnel as well as very precise equipment. For example, test specimen preparation (e.g., by sawing) may already introduce fine cracks in the samples that will negatively affect tensile strength. The exact positioning of the test specimen in the testing machine (fiber orientation and load direction) is also of crucial importance. The testing methods to determine mechanical properties of fiber reinforced materials and a detailed description of specimen preparation can be found in the literature [9].

1.2.2 Manufacturing

The manufacturing methods for fiber reinforced materials are largely dependent on the matrix. Thermoplastic matrices are generally heated, melted, and injected or pressed into a mold. Here, the matrix solidifies when the component cools in the mold. The final geometry of the component is therefore determined by the mold.

On the other hand, resin systems (uncured thermosets) are chemically reactive systems and consist of low-molecular compounds. Similar to thermoplastic materials, resins are placed in a mold or a press. However, the solidification of the matrix is not a result of cooling (as for thermoplastics), but of a cross-linking reaction (curing). As soon as a stable, three-dimensional structure has been formed, the component can be de-molded. Because of the necessary chemical reaction, the time to solidification of a fiber reinforced thermoset is typically longer than for a fiber reinforced thermoplastic material. Therefore, cycle times for thermoset matrices are typically longer. In addition, cross-linking is a non-reversible process, i.e., it is not possible to re-mold a thermoset component after it is cured.

Resins exhibit significantly lower viscosities than molten thermoplastics. Therefore, the flowability of resins is higher, allowing for the manufacturing of more complex component geometries, because thinner walls and longer flow paths are easier to wet. In addition, thermoset fiber reinforced materials require lower process pressures. Some processes, e.g., resin infusion for the manufacture of rotor blades for wind turbines, require no pressure. Instead, the resin is drawn into the mold by negative pressure (vacuum). The design of molds and processing equipment/plants for the manufacture of thermoset fiber reinforced materials is also less involved, lowering capital costs. Therefore, continuous fiber reinforced components are commonly made with thermosetting matrices. The most relevant processes for the manufacture of continuous fiber reinforced thermoset components are listed in Table 1.3.

Table 1.3 Comparison of Different Manufacturing Methods

Method	Personnel cost	Capital investment cost	Level of automation	Production rate	Component quality/ Mechanical properties	Miscellaneous	Application example
Hand lay-up	3	0	0	0	0	Open process (health hazards)	▪ Prototypes and exclusive items ▪ Swimming pools
Vacuum infusion	3	0	0	0	1	Suitable for large parts	▪ Rotor blades for wind turbines ▪ Boats
Pultrusion	0	3	3	3	3	Only suitable for profiles	▪ Pipes, tubes and profiles
Filament winding	0	3	3	1	3	Not suitable for complex geometries	▪ Pressure tanks
RTM	1	2	2–3	2	2	Suitable for mass production	▪ Automotive parts ▪ Small aircraft parts
Prepreg/ Autoclave	2	2	1–2	1	3	Limited prepreg shelf-life	▪ Racing car parts ▪ Aerospace components ▪ Repair patches

0: Low 1: middle 2: high 3: very high

In order to manufacture components with specified properties at desired costs, it is imperative to consider simultaneously materials, process, and component design. These three aspects are closely related. For example, a new component design may require a change in manufacturing process, or an adjustment of the resin system, and/or in fiber orientation in order to achieve optimal component performance.

1.2.3 Applications

Fiber reinforced materials are used in a variety of applications that require very low-weight materials with particularly good mechanical properties. This includes the energy sector (wind turbines), the aerospace industry, the military, and the automotive industry [10]. Over the last years, fiber reinforced materials, in particular carbon fiber reinforced plastics, have seen annual growth rates of approx. 13 %, and they are predicted to increase at a significantly higher rate by 2020.

The reasons for this growth include increasing environmental concerns, both in the general public and within governing bodies, as well as the desire to significantly reduce fuel consumption, in particular in transportation and shipping [11]. Reducing the weight of airplanes and cars/trucks offers the highest potential to lower fuel consumption and CO_2 emissions, thus forcing manufacturers to explore material alternatives with better specific mechanical properties. The use of carbon-fiber reinforced plastics (CFRP) may reduce component weight by as much as 70 % compared to steel, and by as much as 30 % compared to aluminum assuming load-appropriate design [12].

CFRP components are typically manufactured using prepreg technology. More than 54 % of the global carbon fiber production is used for the manufacture of prepregs. Another approx. 5 % is used as rovings or meshes for infiltration processes such as resin transfer molding (RTM). Other important manufacturing processes for CFRP include winding processes (approx. 15 %) and pultrusion (approx. 8 %) [10].

Approximately 95 % of carbon fibers are used in carbon fiber reinforced plastic components. Therefore, carbon fiber consumption is a reflection of the engagement in the field of carbon fiber reinforced components in a certain country or region. Table 1.4 shows the carbon fiber consumption by region and application. Carbon fiber reinforced components for aviation, automotive, and wind turbine applications are primarily produced in Europe, while mass-produced CFRP components for sporting goods are almost exclusively produced in Asia.

Table 1.4 Consumption of Carbon Fibers by Markets and Regions [13]

	Europe	North America	Asia	Other
Aerospace and defense	46 %	33 %	16 %	5 %
Sport and hobby	4 %	6 %	89 %	1 %
Wind energy generators	74 %	15 %	11 %	< 1 %
Automobile	56 %	26 %	18 %	< 1 %

1.3 References

[1] USGS. *U.S. Geological Survey, Minerals Information*. (Online) U.S. Department of the Interior. http://minerals.usgs.gov/minerals/pubs/commodity/ (retrieved June 9, 2014).

[2] ISPY Publishing: *Glass Market Intelligence Report*, Bolton: ISPY Publishing, 2013.

[3] Ceramic Industry Advanced: *Raw & Manufactured Materials: 2014 Overview*. (Online) (accessed 01-02-2014) http://www.ceramicindustry.com/articles/93677-raw-manufactured-materials-2014-overview (retrieved June 9, 2014).

[4] Cerame Unie – The European Ceramic Industry Association: *Activity Report*. Brussels: Cerame Unie, 2013.

[5] Plastics Europe: *Plastics – the Facts 2013, An analysis of European latest plastics production, demand and waste data*. Brussels: Plastics Europe, Association of Plastics Manufacturers, 2013.

[6] G.W. Ehrenstein. *Polymeric Materials – Structure, Properties, Applications*. Munich: Hanser Publishers, 2001.

[7] H. Schürmann. *Konstruieren mit Faser-Kunststoff-Verbunden*. VDI-Buch. Berlin, Heidelberg: Springer Verlag, 2007.

[8] G.W. Ehrenstein. *Faserverbund-Kunststoffe – Werkstoffe – Verarbeitung – Eigenschaften*. Munich: Hanser Publishers, 2006.

[9] V. Altstädt. *Testing of Composite Materials*, in *Polymer Testing*, W. Grellmann, S. Seidler (Eds.). Munich: Hanser Publishers, 2007, 511–560.

[10] E. Witten, B. Jahn. *Composites-Marktbericht 2013 – Marktentwicklungen, Trends, Ausblicke und Herausforderungen*. AVK Industrievereinigung verstärkte Kunststoffe/Carbon Composites, 2013.

[11] C.-S. Ernst, L. Eckstein, I. Olschewski. *CO_2-Reduzierungspotenziale bei Pkw's bis 2020*. Aachen: Final report 113510, 2012.

[12] F. Gojny. *Carbon Fibres & Composites: Ascent to Industrial Engineering Material – Market Outlook & Applications*, in *Carbonfasern: Herstellung – Technische Möglichkeiten – Marktpotenziale*. Meitingen: Cluster-Treff Bayern Innovativ, 2011.

[13] Industry Experts. *Carbon Fibers & Carbon Reinforced Plastics (CFRP) – A Global Market*. 2013.

2 Prepregs and Their Precursors

Felipe Wolff-Fabris, Hauke Lengsfeld, and Johannes Krämer

Over the last 70 years, prepregs have significantly influenced the technological development of high-performance fiber reinforced components. Today, these materials are globally prevalent and are used for the manufacture of composite parts in the aerospace industry, for high-speed trains, cars, boats, and many other applications. More than half of the global carbon fiber production is used to manufacture prepregs [1].

Yet, there are still segments in industry that are discovering the advantages of fiber reinforced composite materials over those of conventional materials, such as metals. Increasingly, combinations of metals and fiber reinforced plastics – so-called hybrids – are introduced.

This chapter will provide an overview of the design of prepregs on the materials typically used to manufacture them. In addition, we will introduce examples of the different generations of prepregs, including the current, modern prepreg systems.

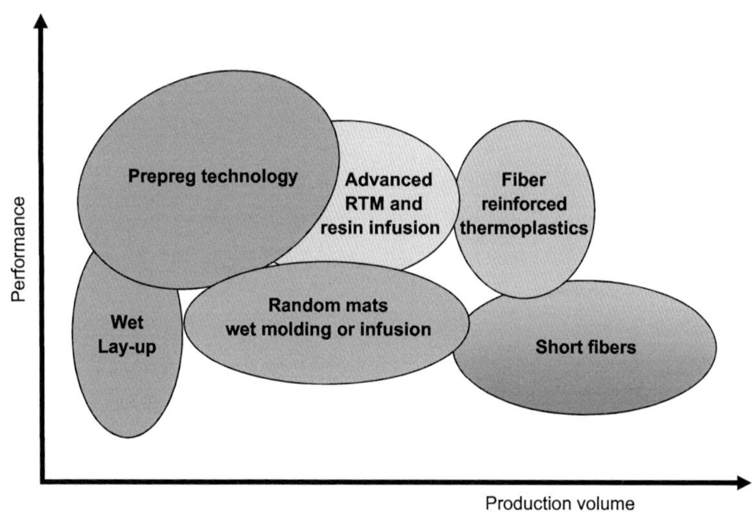

Figure 2.1 Performance and production volume of prepregs compared with other technologies

With regards to performance, prepregs claim the undisputed leading position among high-performance composite materials (Figure 2.1). This class of materials allows the manufacture of ultra-lightweight yet highly load-bearing composite components. However, the figure also shows that the advantage in performance comes at the cost of productivity (see also Chapters 4 and 5). Therefore, this technology is mainly used for very small to medium-sized production runs. In these scenarios, component performance and production cost can be easily reconciled using prepreg materials (Figure 2.2). Despite their lower productivity compared to infusion techniques, prepregs excel in terms of their simple and safe processibility.

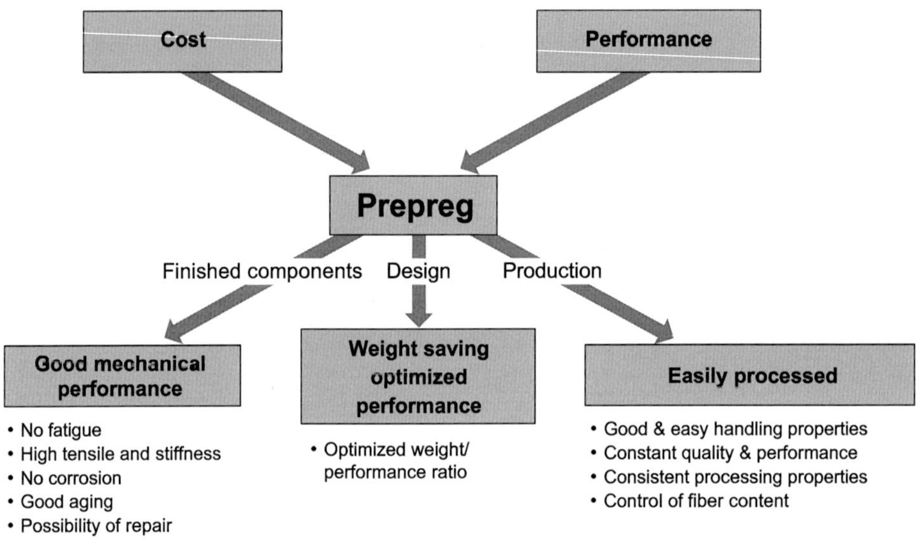

Figure 2.2 Prepregs as key materials for high-performance components

The costs of prepreg materials are determined by the requirements the finished component has to meet. Prepreg costs for the production of aviation components cannot be compared to those for components for wind turbines. The latter application is a good example for the fact that manufacturing and labor costs can be lowered by utilizing prepregs with high areal weight rather than using infusion technology. Using prepregs with areal weights of up to 2000 g/m^2 allows for high lay-up efficiency (even by hand) and faster completion of the component. While the matrix is already part of the prepreg and only has to be cured, dry semi-finished products require the additional step of infusion with the matrix resin.

2.1 Structure and Preparation

A prepreg (*pre*-im*preg*nated fibers) is a semi-finished product consisting of a highly viscous matrix and continuous reinforcing fibers. The material is typically wound on rolls and comes in two variations:

- Unidirectional (UD) prepreg (fiber reinforcement in only one direction)
- Fabric prepreg (orthogonal fiber reinforcement)

In order to manufacture a prepreg, the fibers are wetted with the matrix material, resulting in a pre-impregnated semi-finished product (see also Chapter 3). The choice of matrix material includes thermosetting resin systems as well as thermoplastics; however, thermosetting systems are more commonly used.

The cross section in fgure 2.3 shows the typical composition of a prepreg as it is delivered between a polyethylene film (PE film) and a release paper, ready for further processing. Some processing techniques, such as automatic tape laying (ATL, see Chapter 4), require a carrier material only on one side of the prepreg. Both the PE film and the release paper protect the prepreg from contamination, prevent sticking, and facilitate handling, e. g., during cutting.

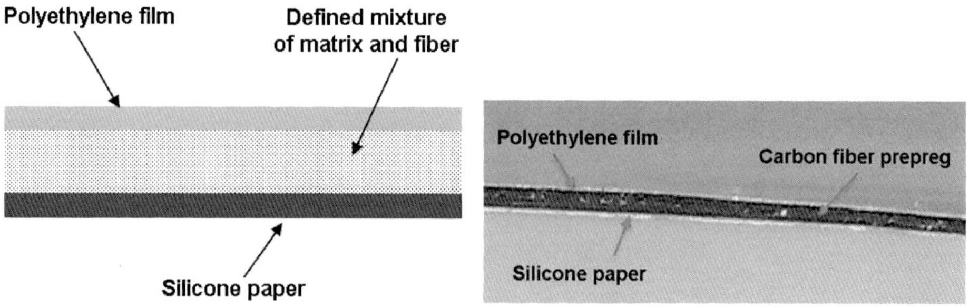

Figure 2.3 Typical prepreg cross section

The dimensions of the semi-finished product vary depending on application and type of material (UD or fabric prepreg). The roll width also depends on the width of the production line. Common roll widths and lengths are shown in Table 2.1.

Table 2.1 Examples of Prepreg Dimensions and Roll Weights

	UD prepreg	Fabric prepreg
Width in mm	75 to 1500	900 to 1500
Length in m	up to 800	up to 200
Roll weight in kg	up to 80	up to 200

The ratio of fiber to matrix content is precisely defined for each prepreg, with narrow tolerances that depend on application. The tolerances for nominal resin content (RC) range from ± 2.5 % to as little as ± 1 %. However, the resin content may be adjustable within certain limits, depending on application-specific requirements. A common choice is a combination of fiber areal weight and resin content that allows for a fiber volume content of approx. 60 wt.% after curing of the material. In many cases, resin is lost during curing in vacuum processes, and therefore the resin content in the semi-finished product is often adjusted to a higher value. Depending on resin content in the prepreg and/or in the composite component, the theoretical value of the cured ply thickness (CPT) changes. Excessive resin loss during the curing process may also lead to a smaller CPT than theoretically calculated. On the other hand, if very little resin was lost during vacuum processing, the CPT may be higher than calculated. Table 2.2 shows typical fiber areal weights, resin contents, and theoretical cured ply thicknesses for selected prepreg types with carbon fibers. The table could be expanded to include many other types of reinforcement, such as glass and aramid fibers.

Table 2.2 Examples of Fiber Areal Weight, Resin Content, and Cured Ply Thickness of Prepregs

Prepreg type	Fiber	Fiber areal weight (FAW) in g/m²	Resin content (RC) in wt.%	Cured ply thickness (CPT) in mm	Fiber volume content (FVC) in wt.%
UD standard Prepreg	Carbon fiber	134	35	0.130	58
	Carbon fiber	134	40	0.143	52
	Carbon fiber	268	35	0.260	57
	Carbon fiber	268	40	0.286	52
Fabric standard prepreg	Carbon fiber	370	35	0.36	57
UD zero-bleed prepreg	Carbon fiber	134	34	0.129	58
	Carbon fiber	268	34	0.258	58

In so-called "zero-bleed" prepregs of the 3rd and 4th generation, the resin content is so precisely adjusted to the fiber areal weight that there is virtually no resin flow (bleeding of excessive resin) during the curing process.

Another characteristic of prepregs is their limited shelf life because of their typically reactive resin matrix. Many prepregs have to be stored at −18 °C in order to decelerate the crosslinking reaction. With appropriate cooling, shelf lives between 6 and 18 months can be achieved. Lately, prepregs with up to 60 days storage stability at room temperature have been commercially available for general industrial and automotive applications. Prepregs based on thermoplastic matrix systems do not undergo detrimental ageing and can be stored indefinitely.

2.2 Matrix Systems

Both thermoplastics and thermosets offer suitable matrix systems for prepregs; they are commonly called thermoplastic organo sheets. Thermoplastic prepregs based on polypropylene (PP) and polyamide (PA) offer relatively little temperature resistance and are therefore mostly used in the automotive industry. In aviation, where temperature resistance up to 180 °C is commonly required, high-performance polymers such as polyether sulfone (PES) or polyether ether ketone (PEEK) are typically used. For example, components can be industrially manufactured from thermoplastic fiber reinforced plastic using fiber forge technology.

Thermoplastic prepregs do not undergo a chemical reaction or crosslinking during processing, because the matrix material is already completely polymerized. With their almost indefinite storage stability and shorter component manufacturing cycles, thermoplastic prepregs offer advantages over thermosetting prepregs. An additional advantage of the thermoplastic matrix is its very high impact strength.

Nonetheless, thermoplastic prepregs are used in only few applications. This can be attributed to the high viscosity exhibited by thermoplastic materials, which adversely affects consolidation and air removal. As a result, thermoplastic fiber reinforced components may exhibit significantly higher porosity and correspondingly more defects than components made from thermosetting prepregs. In addition, thermosetting prepregs are exclusively processed in presses, allowing for only simple and small component geometries. Continuous fiber reinforced components in aviation applications are therefore typically produced using thermosetting matrices.

Thermosets are characterized by the fact that their final molecular structure is developed only during the manufacture of the component. This process is called crosslinking, crosslinking reaction, or curing. The developing three-dimensional structure can be influenced by curing temperature and time and also during post-curing.

Generally, thermosets are polymerized by the reaction between at least two low-molecular materials, the resin and the curing agent. The increase in molecular weight created by the reaction of the two starting materials (reactants) causes an increase in viscosity (Figure 2.4). The final property profile of the component is critically influenced by the reactants used and by the curing conditions (see also Chapter 5).

The following provides a short description of the most important thermosetting resin systems for prepreg applications.

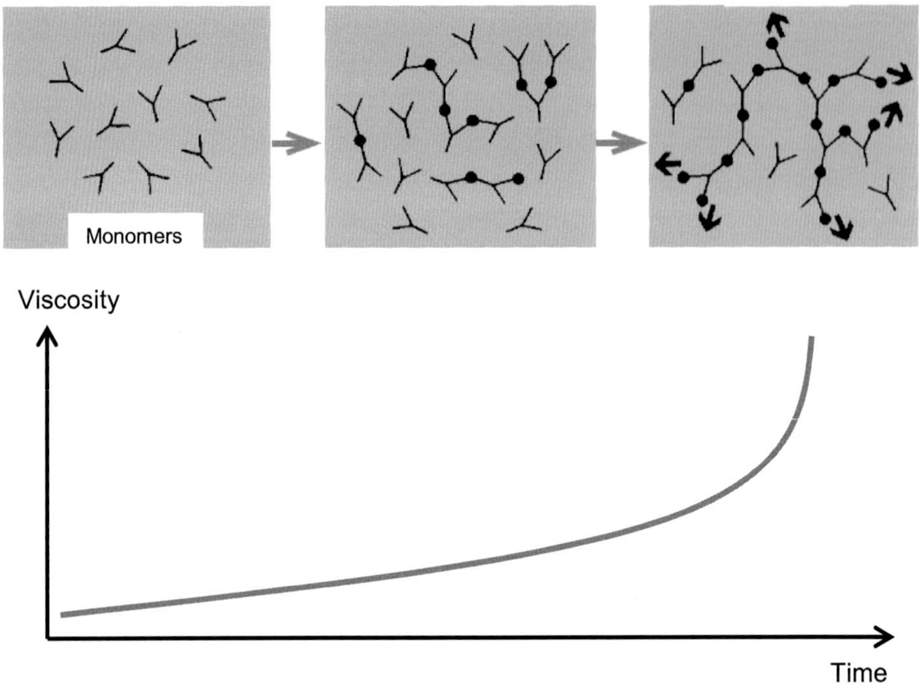

Figure 2.4 Schematic curing characteristics of a thermosetting material

Epoxy resins are the primary class of thermosets used for prepregs. They form cross-linked systems by a polyaddition reaction of monomers with an epoxy group and suitable curing agents. Commonly used epoxy monomers include *N,N,N',N'*-tetraglycidyl-4,4'-diaminodiphenylmethane (TGDDM or TGMDA), diglycidyl ether of bisphenol A (DGEBA), and glycidylated novolacs (epoxy novolacs). The degree of crosslinking in epoxy resins strongly depends on the ratio between epoxy resin monomer and curing agent. Maximum crosslinking density is reached at a stoichiometric ratio, at which theoretically each reactive group forms a crosslinking point. However, the state of complete cure is in fact never reached, because the perfect ratio is metrologically not obtainable. In addition, towards the end of the curing process the reduced chain mobility results in a steric hindrance, thus preventing the completion of the curing reaction.

Epoxy resins can be cross-linked with a variety of curing agents; here, amines are the most commonly group utilized. Typically, aromatic amines, such as 4,4'-diaminodiphenyl sulfone (4,4'-DDS), dicyandiamide (DICY), or anhydrides are used as curing agents for prepregs. These systems exhibit relatively long shelf lives and are cured at high temperature (commonly 180 °C) over several hours. The addition of accelerators or the substitution of aromatic by cyclic aliphatic amines facilitates the curing of prepregs at lower temperatures (e.g., 120 °C). Linear aliphatic amines are not suitable as curing agents for prepreg epoxy resins, because here the crosslinking reaction already starts at room temperature so they do not offer the required shelf life.

Bismaleimide resins (BMI resins) are cross-linked using aromatic amines or thiols and are generally cured at higher temperatures than epoxy resin systems. In order to take full advantage of the property profile of BMI resins, components are post-cured at temperatures above 200 °C. Typically, BMI resins find applications when the thermal resistance offered by epoxy systems is not sufficient.

Phenolic resins provide very good fire/smoke toxicity (FST) characteristics and are often the matrix of choice for fiber reinforced components in interior applications, such as in linings and trim for automotive, mass transit, rail car, and aerospace interiors. Because of their low toughness and strength, phenolic resins are rarely used in structural applications. They are also more demanding during processing because their polymerization reactions form volatile compounds. Components made from phenolic resins therefore often exhibit higher porosity than those made from epoxy or BMI resins prepregs.

Benzoxazine resins (BOX resins) represent the next step of development of phenolic resins. These resins do not form any volatiles during curing, making the production of high-quality components possible. The flammability properties of BOX resins are almost as good as those of phenolic resins. At room temperature, it is basically impossible to initiate a curing reaction, which allows for safe transport and storage lives exceeding 12 months. BOX resins typically exhibit higher viscosities than other thermosets, a fact that has to be considered during the selection of curing parameters.

Table 2.3 Properties and Applications of Various Resin Systems

Matrix	Properties	Application examples
Epoxy resin	▪ Very high mechanical properties ▪ Easy processing	Sports goods, automotive parts, rotor blades for wind energy, wing cover and fuselage parts for aircraft
Bismaleimide resin	▪ Very high thermal stability	Aircraft engines
Phenolic resin	▪ Very good FST properties	Interior parts for trains and aircraft cabins
Benzoxazine resin	▪ Very high storage stability ▪ Good FST properties	Potential applications in aerospace (not yet established)

2.3 Fibers and Textiles

Theoretically, there is a wide variety of different fibers with different compositions and properties; however, the most commonly utilized types are glass and carbon fibers. Standardized prefixes provide a more differentiated classification, for example E-glass (electrical) or HM carbon fiber (high modulus).

Figure 2.5 shows a comparison of the properties of selected fibers. The values listed here are reference values. We assume some general knowledge, such as that thinner fibers exhibit higher strength values than thicker fibers of the same material and that the measured strength of an individual fiber increases with decreasing clamping length.

The most commonly used type of fiber is E-glass. Higher strength values, such as those seen in S-glass (strength) and R-glass (resistance), can be achieved by variations in composition.

HS glass fibers (high strength) are common reinforcing materials with good characteristics. HM fibers exhibit higher Young's modulus values, while IM fibers (intermediate modulus) offer higher strength. Both glass and carbon fibers are offered in additional, special types.

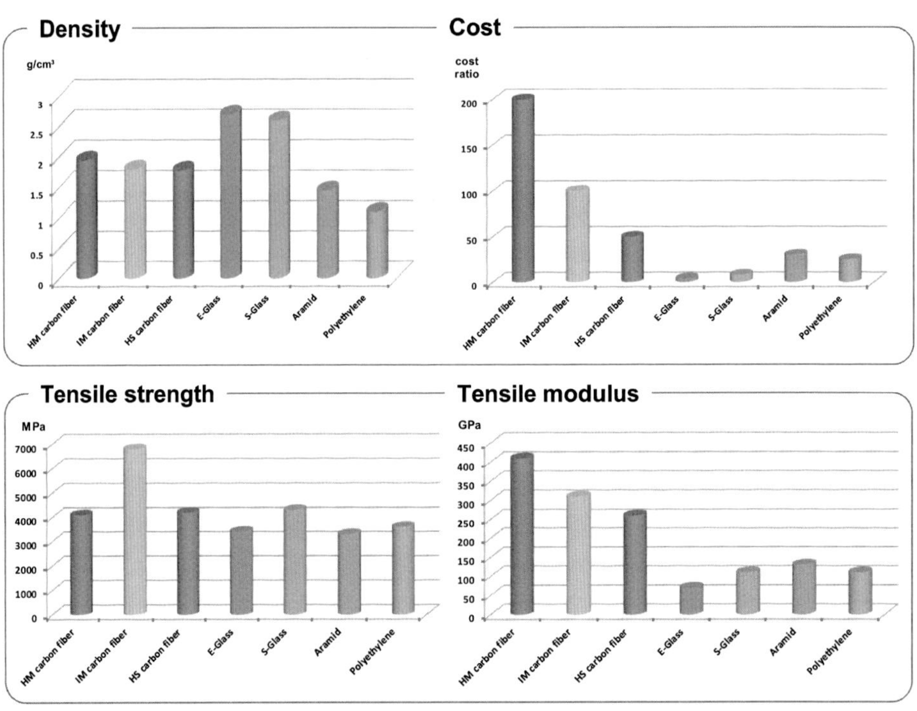

Figure 2.5 Properties of various fibers

Multifilaments can be directly manufactured into unidirectional prepregs; in addition, textile semi-finished products, such as *fabrics or multi-axial (non-crimp) fabrics*, can be used. However, the latter require an intermediate textile production step prior to the manufacturing of prepregs.

In woven fabrics, the reinforcing fibers are manufactured into planar semi-finished products using conventional weaving techniques. Figure 2.6 (top) shows common weave patterns. The choice of weave pattern has significant influence on subsequent processing and the final mechanical properties of the component. For example, a plain weave is easier to handle than a twill weave. However, the high number of interstices in the plain weave causes strong fiber undulation and thus leads to a reduced property profile. Reducing the number of weft fibers in the satin weave creates a quasi UD fabric with more than 80 % fiber content in the warp direction.

Figure 2.6 Schematic of different weave patterns

In contrast to woven fabrics, non-crimp fabrics (NCF) do not exhibit fiber undulation and can be arranged in a variety of angles. The different plies are held in place by adhesive layers or by stitch bonding. Figure 2.7 shows a typical quadraxial ply stack.

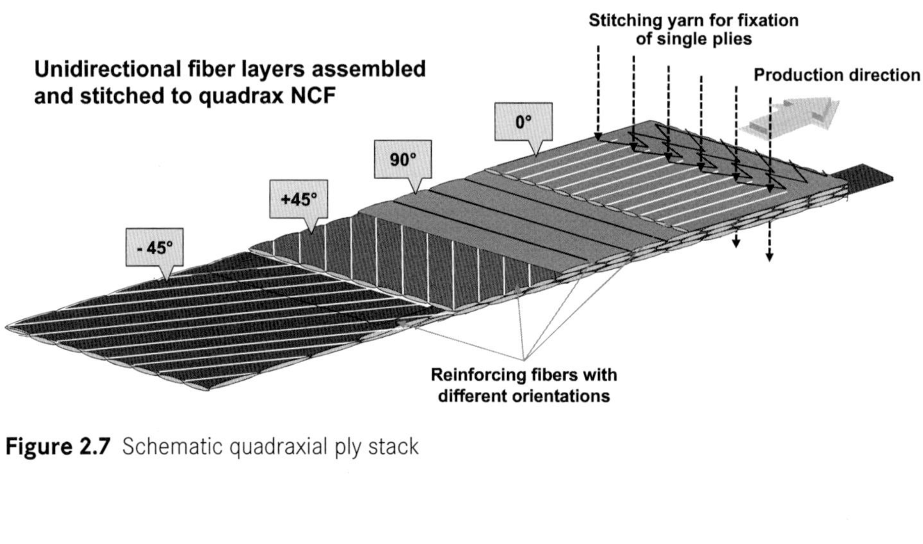

Figure 2.7 Schematic quadraxial ply stack

Table 2.4 provides a short overview of the properties and applications of different types of reinforcement. More detailed descriptions of fibers and semi-finished products can be found in the literature [3–6].

Table 2.4 Properties and Applications of Different Types of Reinforcement

Reinforcement	Properties	Applications
UD prepreg tape	▪ Variable fiber areal weights from 100 g/m² to 800 g/m² (carbon), or 3000 g/m² (glass) ▪ High strength and stiffness in one direction	Sports goods, aircraft primary structures
Slit tape/tow preg	▪ High strength and stiffness in one direction ▪ Suitable for fiber placement on complex 3D geometries ▪ High strength and stiffness in one direction ▪ Very narrow width for accurate fiber placement (1 mm)	Aerospace, pressure vessels, drive shafts, tubes
UD fabric (> 80 % warp)	▪ High strength and stiffness mainly in one direction ▪ For components requiring predominant strength and stiffness in one direction	Aerospace, industrial, sport and leisure
Balanced woven fabrics	▪ High strength and stiffness in two directions ▪ Very good handling characteristics and good drape ▪ Choice of weave styles ▪ Possible to mix fibers, hybrid fabrics possible ▪ Weights from 20 to 1,000 g/m²	Aerospace, industrial, sport and leisure
NCF	▪ Time-saving, cost-effective technology ▪ Strength and stiffness in multiple directions ▪ No crimp ▪ Ability to optimize weight distribution in fabric	Large advanced structuresAerospace, e.g. floor beams Industrial Wind energy

2.4 Prepreg Systems

Their chronological technological advancement allows for prepregs to be categorized into four generations from a historical perspective:

- *1st Generation:* Fiber material with resin system (resin + curing agent)
- *2nd Generation:* Fiber material with resin system + dissolved impact modifier (so-called toughener)
- *3rd Generation:* Fiber material with resin system + dissolved impact modifier + insoluble thermoplastic particles at the surface of the prepreg to further increase impact resistance (so-called interleaf prepregs)
- *4th Generation:* Interleaf prepreg with improved electrical conductivity, e.g., against lightning strikes (some still under development)

The first generation provides more or less the blueprint for prepregs and consists of highly viscous resin systems; however, they are relatively brittle and exhibit only limited fracture toughness. In order to improve this property, a toughener was added to the resin system, in particular to epoxy systems. Depending on the specific application, rubbers or thermoplastic materials were used as tougheners. Today, 1st generation prepregs are often used in interior applications in the aircraft industry. In the course of development it became clear that in particular high-temperature thermoplastics, such as polyether sulfone (PES) and polyether imide (PEI), do not negatively affect the Young's modulus of the resin system, an effect commonly seen with rubbers. Therefore, prepregs of the 2nd generation are prepared using finely ground PES which is added and dissolved in the epoxy resin. These prepreg types are still commonly used and have been successfully implemented in aircraft applications, such as horizontal and vertical stabilizers, for approx. 25 years.

With the development of other aircraft parts from carbon fiber plastics, such as wings and fuselages, the impact strength requirements increased and were subsequently met by prepregs of the 3rd generation, the interleaf prepregs, which emerged in the 1990s. Here, a very thin layer of thermoplastic material is applied to the surface. This thermoplastic material has a specified particle size and is insoluble in the resin system (Figure 2.8). These particles create a uniform distance between the carbon fiber tows in the component, where potential cracks can be arrested either by the particles or by a so-called hackling mechanism [2, 3].

Figure 2.8 provides a comparison between prepregs of the 2nd and 3rd generation. It clearly shows the increased distance between the carbon fiber tows in the interleaf prepreg compared to the 2nd generation prepreg. The surface of interleaf prepregs often appears milky gray because of the added fine thermoplastic powder.

In order to further optimize interleaf prepregs for aviation applications, in particular fuselage applications, the electrical properties in the through-thickness direction of 4th generation prepregs required substantial improvement, without compromising other properties. This tailored improvement of individual properties is one of the biggest challenges in current prepreg development.

The number of currently commercially available prepreg systems is so high that it is basically impossible to provide a comprehensive overview. The large variety can be attributed to the wide bandwidth of requirements met by prepreg materials in the different applications, ranging from aviation to general industrial applications.

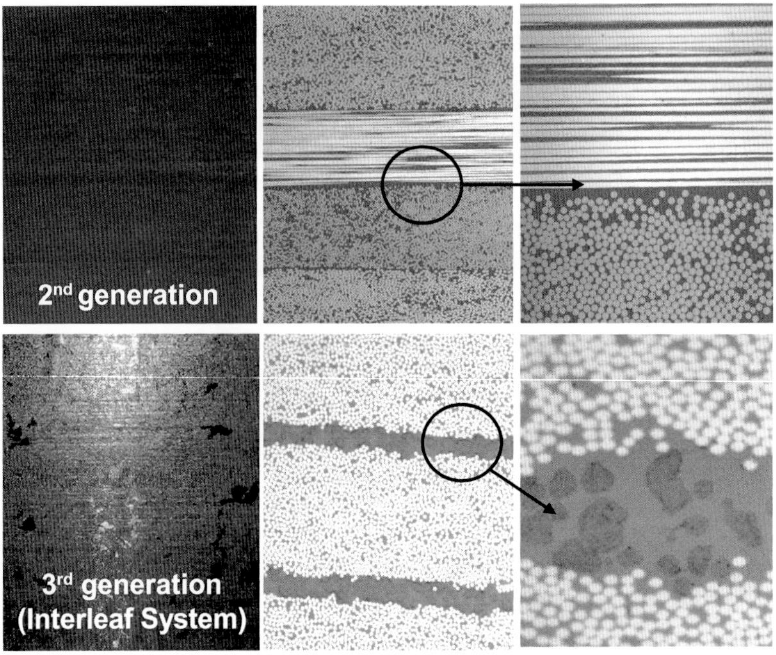

Figure 2.8 Surface and polished micrograph section of 2nd and 3rd generation prepregs

Table 2.5 illustrates a small section of the prepreg systems currently used in industrial applications. Because of their long history, versatile properties, and ready availability, epoxy-based systems have gained wide acceptance and application. However, systems based on bismaleimide (BMI) and recently also benzoxazine have found applications in which epoxy resins do not meet requirements, e.g., high temperature applications up to approx. 300 °C. This overview could be expanded to include other resin types such as phenolic- and cyanate ester resins, which are typically used in aviation interior and aeronautic applications. The latest systems utilized for civil aviation applications include HexPly® M21E and Toray® 3911, which are used as structural prepregs in A350XWB (Airbus) and B787 (Boeing) airplanes. Applications in the automotive and general industrial sector include prepreg-reinforced B-pillars (hybrid construction of steel sheet and fiber reinforced composite) and rotor blades for wind turbines.

Table 2.5 Selection of Current Prepreg Systems and Manufacturers

Product	Manufacturer	Resin type	Curing temperature (°C)	Glass transition temperature (°C)	Manufacturing process	Curing technology	Application (aero = aerospace)
HexPly® 8552	Hexcel	epoxy	180	180		autoclave	aero, structure
HexPly® 914	Hexcel	epoxy	175	180		autoclave	aero, structure
Cycom® 997	Cytec	epoxy	180	210		autoclave	aero, structure
Cycom® 977-2	Cytec	epoxy	180	126–138		autoclave	aero, structure
Cycom® 977-3	Cytec	epoxy	177	177		autoclave	aero, structure
HexPly® 913	Hexcel	epoxy	125–160	125–160		autoclave	aero, structure, repair
HexPly® M92	Hexcel	epoxy	125	160		autoclave	aero, structure, interior
HexPly® M77	Hexcel	epoxy	80–160	120–160		vacuum, press	industry, automotive, sports
HexPly® M65	Hexcel	BMI	190–230	300		autoclave	aero, structure
Cycom® 5250-4	Cytec	BMI	177–200	200		autoclave	aero, structure
Cycom® 5320-1	Cytec	epoxy	177	177	hand lay-up ATL AFP	out-of-autoclave	aero, structure
HexPly® M56	Hexcel	epoxy	180	190–220		out-of-autoclave	aero, structure
MTM44-1	Cytec	epoxy	130	180		out-of-autoclave	aero, structure
MTM55	Cytec	epoxy	120	120		autoclave, out-of-autoclave, press	industry, automotive
HexPly® M21	Hexcel	epoxy	180	180		autoclave	aero, structure
HexPly® M21E	Hexcel	epoxy	180	180		autoclave	aero, structure
HexPly® M91	Hexcel	epoxy	180	185–190		autoclave	aero, industry, structure
Toray® 3911	Toray	epoxy	180	?		autoclave	aero, structure
LOCTITE®BZ9703	Henkel	benzoxazine	177	120		autoclave	aero, structure
LOCTITE®BZ9704	Henkel	benzoxazine	185	140		autoclave	aero, structure
LOCTITE®BZ9705	Henkel	benzoxazine	210	180		autoclave	aero, structure
EP112/EH250	Gurit	epoxy	120–160	110	hand lay-up	autoclave, press	aero, sec. structure
PB1000	Gurit	benzoxazine	140–160	140–160	hand lay-up	autoclave, press	aero, interior, sandwich
SIGRAFIL® types	SGL	epoxy	130–170	120–180	hand lay-up	autoclave, press	industry, automotive

Other than by their mechanical properties, prepreg systems can also be distinguished by their curing temperature. This characteristic has not only decisive influence on the maximum service temperature of the cured component, but also on the required curing technology (Chapter 5). At the same time it represents the huge versatility of the available prepreg systems. Over time, two curing temperature ranges (approx. 120 to 140 °C and 180 °C) have emerged, which can be attributed to the curing agents available for epoxy-based systems. For specific high-temperature applications, components are cured at temperatures up to 240 °C (e.g., BMI). Typically, all prepregs can be used both with automated manufacturing techniques (ATL, AFP: see Chapter 4) as well as with hand lay-up methods.

Because of their internal "architecture", many prepregs have to be cured under pressure in order to produce non-porous parts and to optimally exploit the properties offered by the material. Aviation applications therefore often require autoclave curing. Because of the associated high capital and operational costs, there is a trend to manufacture even structural parts using out-of-autoclave (OOA) curing technologies. Here, it is possible to cure the prepregs using only vacuum, without additional pressure, into non-porous components. Prepregs for industrial applications are typically made to allow for either vacuum or hot press curing.

Another current development trend explores dry-fiber-placement (DFP) technology, in which dry carbon fiber rovings are placed in a mold via AFP and then *preconsolidated.* In a subsequent step, the rovings are wetted with liquid resin and cured. The major advantage of DFP technologies versus traditional prepreg technologies is the fact that the semi-finished material can be stored almost indefinitely at room temperature. Prepregs on the other hand typically require cool storage because of their reactive resin systems, and therefore they have limited shelf life. Current products commercially available include HiTape® by Hexcel Composites and PRISM TX1100® by Cytec. The internal architecture of these systems allows for mechanical properties matching those of interleaf prepregs, thus closing the existing property gap between prepreg and infusion systems.

Even though prepreg technologies have been for years declared outdated, market development and the number and varied topics of research publications in this area show the opposite. Unmatched by other materials, prepregs allow for lightweight construction options that can otherwise not be realized.

2.5 References

[1] B. Jahn, E. Witten. *Composites-Marktbericht 2013*. AVK, CCeV, 2013.

[2] V. Altstädt. *Effect of the Polymer Matrix on the Properties of Advanced Composites*. Makromol. Symp. 1991, 50, 137–145

[3] V. Altstädt, D. Gerth, H. G. Recker, M. Stängle. *Interlaminar crack growth in third generation prepreg systems*. Polymer. 1993, 34, 907–909.

[4] P. Morgan. *Carbon fibers and their composites*. 1st ed. Taylor & Francis Group, 2005.

[5] F. Fourné. *Synthetic Fibers*. 1st ed. Munich: Hanser Publishers, 2008, 548.

[6] M. Neitzel, P. Mitschang. *Handbuch Verbundwerkstoffe*. 1st ed. Munich: Hanser Publishers, 2004, 57.

[7] M. Neitzel, P. Mitschang. *Handbuch Verbundwerkstoffe*. 1st ed. Munich: Hanser Publishers, 2004, 25.

2.6 Additional Information

- Product specification sheets: Hexcel Composites, 2014.
- Product specification sheets: Cytec Engineered Materials, 2014.
- Product specification sheets: Henkel Loctite. Benzoxazine Resin Technology. 2013.
- Product specification sheet: Gurit AG. Benzoxazine Resin. 2014.
- Product specification sheet: Gurit AG. Epoxy Prepreg. 2014.
- Product specification sheet: SGL Technologies GmbH. High-Performance Prepregs. 2014.
- Prepreg Technology. Hexcel Composites. 2012.

3 Prepreg Technology

Hauke Lengsfeld and Mike Turner

This chapter will provide an overview of the historical development of prepreg technology. We will detail the currently most important technologies used for manufacturing prepreg materials on an industrial scale as well as a number of different prepreg variations.

3.1 Historical Background

Most of those involved in the composite manufacturing industry are familiar with the development and industrialization of prepreg technology.

In the early 1940s, Dr. Norman De Bruyne developed the general principle of fiber reinforced composite materials for structural applications (he founded Aero Research Ltd in 1934 for this purpose). The material he developed was based on flax fibers and a phenol-formaldehyde resin as the matrix. This material was placed in molds in long segments, pressed, and cured. The segments were subsequently mechanically cut and used as the very first fiber reinforced composite components. At the end of World War II, this technique was used to produce flax-fiber reinforced structural fuselage parts for "Spitfire" airplanes. De Bruyne was convinced that suitably adapted versions of his materials could be used for the production of bigger airplanes. His fiber reinforced composites, made from thermosetting resins and oriented fibers, provided the required properties in terms of strength, stiffness, and fatigue resistance.

In 1963, Leslie Phillips and his team at the Royal Aircraft Establishment (RAE), Farnborough, UK, were pioneering pre-impregnated fibers. They developed the first carbon fiber in the form of a multi-filament tow impregnated with a resin matrix that could be used to manufacture structural components. It represented the concept of unidirectional prepregs (UD prepreg), or pre-impregnated sheets.

Based on the developments made in Farnborough, the British company Courtaulds was one of the first to scale up the manufacturing process to make carbon fibers as tows of filaments (fiber types Grafil XA-S and Grafil HM-S), suitable as reinforcement in prepregs.

In the early 1970s, CIBA (Bonded Structures Ltd.) in Duxford Cambridge, UK, used the results of De Bruyne's work on pre-impregnated materials and developed a prepreg with carbon fibers made by Courtaulds and epoxy resin. They quickly realized that prepregs required fibers with high parallel orientation and precise control of areal weight of fibers and resin matrix in order to achieve and maintain high property profiles. The goal to develop a prepreg material that was easily manufactured and processed inspired numerous research projects and studies. To this day, research efforts are focused on the improvement and further development of prepreg materials.

The first prepreg machine consisted of a large drum winder on which dry carbon fiber tows were wound and subsequently impregnated in an epoxy resin bath. This relatively low-cost and low-effort prepreg manufacturing method is still used for evaluation and research purposes. At the same time, epoxy resin coatings were successfully introduced as adhesives for metal–metal bonds in the aircraft and other industries. It was a logical step to utilize these resin coatings for the manufacture of high-value UD prepregs. This led to the development of specific laminating/coating machines that fabricated these resin coatings and also allowed the combination of resin film and dry reinforcing fiber to form prepregs with consistent and predictable areal weight (PAW). The high repeatability of the prepreg areal weight, which is determined by resin weight and fiber areal weight (FAW), is an important quality characteristic of high-value prepregs, in particular in the manufacture of composite components for the aircraft industry. Today, this characteristic is controlled by line speed and impregnation temperature and pressure, providing pre-impregnated materials on an industrial scale. Typically, prepreg thickness ranges between 0.127 mm (0.005 ") and 0.254 mm (0.010 ").

It was quickly discovered that it was possible to manufacture light, high-strength fiber composite components by appropriately cutting and layering several prepregs. During the 1970s and 1980s, fabrics were developed as reinforcing materials for prepregs. Here, the components' properties in different directions could be tailored by choosing different weaves, which were made on looms adapted from silk production; these machines were originally used to make silk garments for aristocrats.

The current state-of-the-art prepreg materials have developed in the context of related technologies over the last 50 years. In fact, the number of carbon fiber materials is significantly higher today because of the development of IM and HS fibers. In addition, the use of natural fibers, such as flax, as reinforcing materials has seen renewed interest in research and development projects. Within a short time period, the carbon fiber manufacturing and processing industries have evolved into sophisticated technologies. Starting from a few selected materials and fiber composites, which were produced in very labor intensive, manual processes, today a variety of industries, including the aerospace and automotive industries, rely on low-weight, stiff, high-performance composites that could not be produced if it were not for prepreg materials and fully automated processes.

3.2 Introduction: Manufacturing Methods

Prepregs, as semi-finished products containing pre-impregnated fibers (Figure 3.1), are a prerequisite for high quality and load-optimized lightweight fiber composites. The current manufacturing techniques and processes guarantee consistent quality of the prepregs, e.g., in terms of FAW and resin content, at a very high level. On the one hand, this simplifies processing for the manufacturer, and on the other it allows the reproducible production of high-quality components.

Prepregs always consist of a combination of a typically highly viscous matrix and a fiber reinforcement. Once the reinforcing material has been pre-impregnated with a matrix, it is considered a prepreg material. Both thermoplastic and thermosetting materials (reactive resin systems) can be used as matrices (see also Chapter 2) [2]. In the following, we will discuss the production of thermoset prepregs.

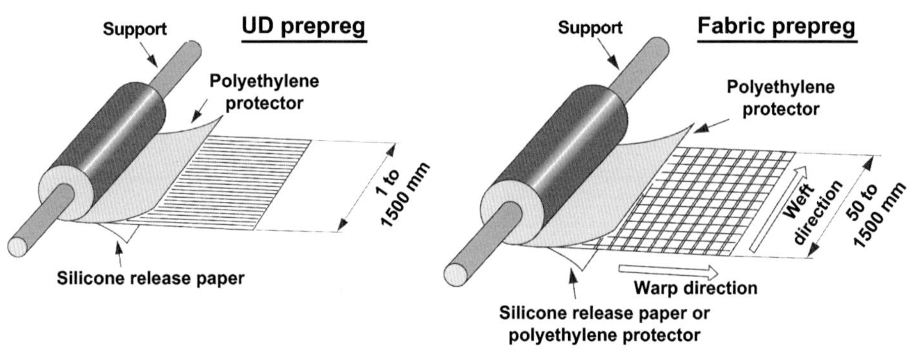

Figure 3.1 Delivery form of UD prepreg (left) and fabric prepreg (right) on supports [Courtesy: © Hexcel Corporation]

Fiber reinforcements are supplied in a number of different forms:

- uni-, bi-, and multidirectional fabrics
- fabrics
- non-wovens and random fiber mats

Depending on the reinforcing structure, processing techniques include creel sets (for rovings) and roll unwinders for fabrics.

Today, a number of different methods and machine concepts are available for the production of coating lines. They are able to use different fiber materials and combine them with a wide selection of matrix resins. State-of-the-art technologies include:

- solution coating processes (also called dipping (solvent) processes) (Figure 3.2)
- hot melt processes (Figure 3.2)
- knife systems
- powder scattering
- slot die systems

Some processes, such as powder scattering, were initially developed for thermoplastic matrices, but later adapted to thermoset processing. However, these processes have not found widespread industrial-scale use. Today, there are two major impregnating methods in industrial practice: the dipping (solvent) process and the hot melt process [2].

Not only do the base materials require cautious handling, but many parameters, including winding, control, impregnating, and drying, have to be considered and adapted individually depending on the type of reinforcement, the impregnating matrix, and the properties of the final product [1].

Thermoset prepregs are available in four typical variations: UD prepreg, fabric prepreg, slit tape, and tow prepregs.

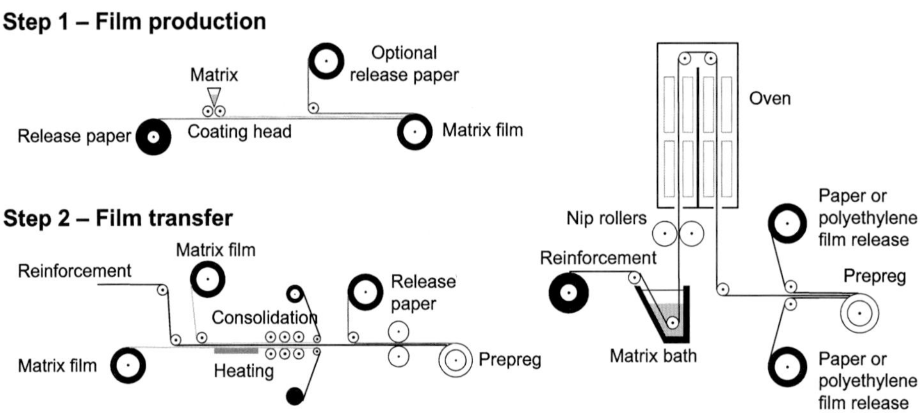

Figure 3.2 Manufacturing methods, left: hot melt process; right: solvent process

3.2.1 Design of a Prepreg Production Line

In principle, a prepreg production line consists of the following stations:

- Creel set for rovings or roll unwinder for fabrics
- Coating unit
- Calendering unit
- Cutting/trimming station
- Winding station for UD or fabric prepreg

Figure 3.3 shows the schematic design of a line producing UD-fiber prepregs.

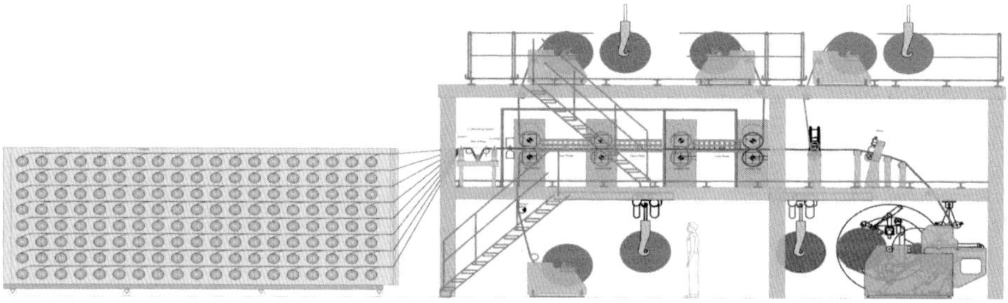

Figure 3.3 Schematic design of a prepreg line [Courtesy: © KROENERT GmbH & Co KG]

Depending on whether the prepreg is manufactured using solvent or hot melt processes, the design of the prepreg line is process- or material-specific.

3.2.2 Prepreg Manufacturing Methods

In the following we will provide a more detailed description of design and process cycles for a UD prepreg line using a hot melt process and a fabric prepreg line using a solvent process.

3.2.2.1 Hot Melt Processes

As mentioned earlier in this chapter, hot melt processes require the manufacture of a suitable resin coating (Figure 3.4). To provide these coatings, base materials including epoxy resins (with a viscosity of approx. 20 Pa·s), tougheners (impact modifiers), curing agents, and possibly other components are mixed. Because prepreg resin systems are extremely viscous (a result of the toughener) and will cure at elevated temperatures, these reaction compounds are often homogenized using extruders. The pre-mixed resin system is extruded and then applied to silicone-coated carrier paper by heated calender rolls. Subsequently, the material is rolled on cores and refrigerated in order to slow the curing reaction and keep the material "fresh" (tacky) for the next production step. Until the mid-2000s, resin films were often produced in batch processes (discontinuously). Here, the resin system was often provided in the form of blocks that were either refrigerated for storage or used directly for the production of resin coated film. The blocks were heated in the coating unit and the epoxy resin transferred to the carrier paper. With the increasing use of prepreg materials in the aircraft and other industries, the so-called continuous-mix process was introduced. It allows for the continuous production of large quantities of a particular resin film. This process has the advantage of providing consistent product quality (e.g., in terms of areal weight) even for industrial-scale production runs. Large quantities of resin film can be produced in a central location and provided to manufacturers, typically prepreg manufacturers with close end-user proximity, at much lower logistical cost, while ready-to-use prepregs would require refrigerated transport. The manufacturing chain can be further automated by directly feeding the resin film into a prepreg line. This approach is investigated for mass-production of fiber reinforced components, e.g., for automotive applications.

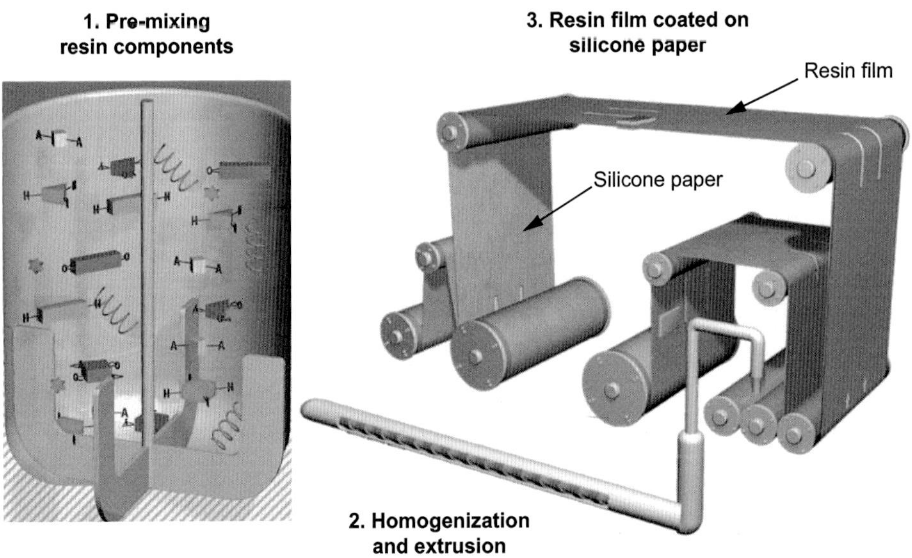

Figure 3.4 Schematic of resin film production [Courtesy: © Hexcel Corporation]

Figure 3.5 schematically shows the sequence of process steps in a hot melt prepreg line. On the left-hand side of the prepreg line is a creel set holding bobbins with glass or carbon fibers. A typical creed may hold several hundred bobbins, feeding into the line from two sides. The number of bobbins is determined by the nominal width of the produced prepreg and its FAW. From each bobbin, a fiber roving or tow is fed into the prepreg machine by guide rolls and fiber eyelets. The fiber tension of each roving or tow is a crucial parameter for high-quality prepregs. In order to set and control fiber tension, each axle on the creel set is equipped with a special brake mechanism that maintains a uniform level of fiber tension on all bobbins. Among other factors, fiber tension influences the tensile strength of the cured material. It also prevents fiber buckling and wrinkles in the finished prepreg.

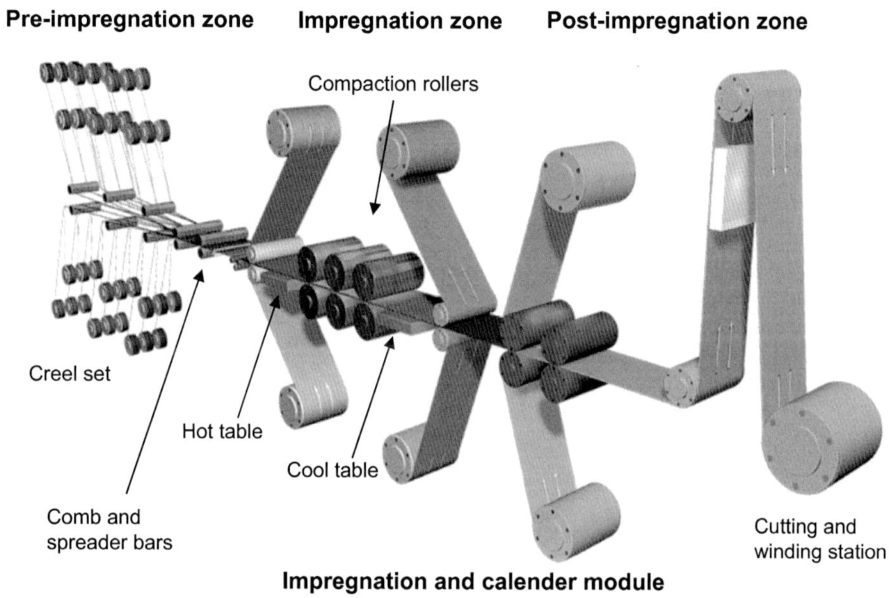

Figure 3.5 Schematic process steps of hot melt systems [Source: © Hexcel Corporation]

Next, the fiber rovings pass through a comb in the pre-impregnation zone (Figure 3.6) that maintains their parallel integrity and prevents crossovers and damage. In addition, this step defines the nominal width of the fiber bed and creates a uniform, closed surface of UD fibers. Simultaneously, the number of rovings per comb hole defines the FAW of the finished prepreg.

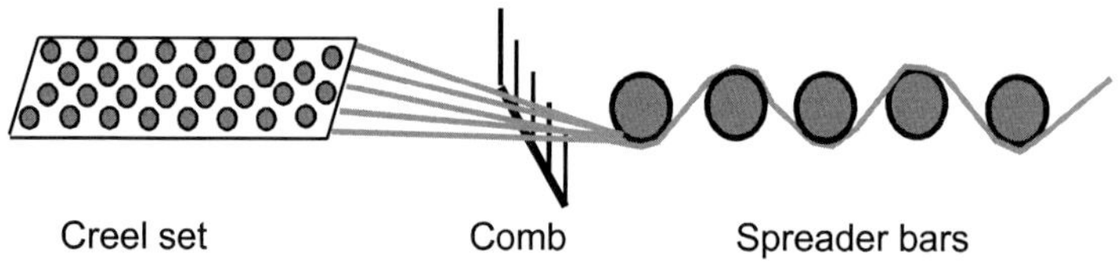

Figure 3.6 Comb and spreader bars, schematic

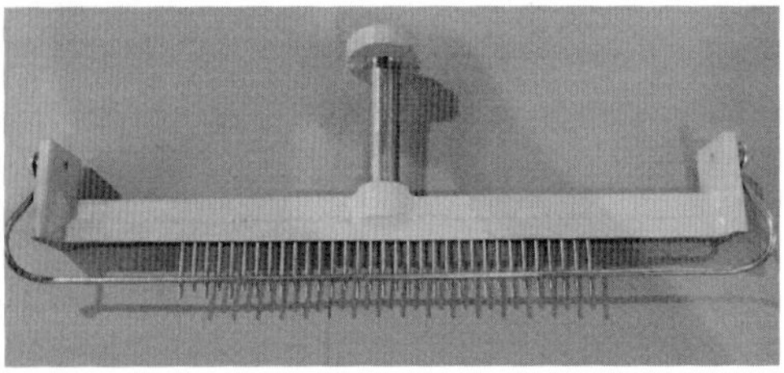

Figure 3.7
Comb in pre-impregnation zone [Courtesy: University Bayreuth, Department of Polymer Engineering, Th. Neumeyer]

Now, the fiber rovings/tows are guided through several spreading and guiding bars in order to achieve the necessary fiber tension to feed them into the impregnation zone. The tension is controlled by the pitch of the guide bars. The spreading bars also help control the FAW over the production width, which is determined by the comb. Depending on how wide the fiber rovings are spread, FAWs of 134 g/m^2 to 268 g/m^2 can be reached in prepregs with 12 k (= 12,000) carbon fibers. The spreading bars also facilitate a homogeneous distribution of fibers across the prepreg area, which is crucial to achieve uniform impregnation during the subsequent impregnation step. Also, a uniform FAW distribution throughout the component is prerequisite for high-performance applications with consistent properties. Fiber and spread also influence the wettability of the fiber bed and thus a variety of other production and processing parameters.

During the next step the fibers are pulled into the impregnation zone (Figure 3.5). Here, the fiber bed is sandwiched between resin films on silicone-coated carrier papers and pulled through the first calender roll (2-film process). The resin content in the uncured prepreg (provided in wt.%) can be controlled by varying the areal weight of the resin films. The areal weight of the two resin films is not necessarily identical. Increasing the resin content on only one side allows enhancement of the tack of the prepreg material. Typical resin contents range from 30 to 45 wt.% for carbon fiber prepregs, and from 40 to 60 wt.% for glass fiber prepregs. In a second impregnation step, additional resin film may be applied (either on one or on both sides) in order to introduce additional properties or to enhance tack (4-film process).

Once passed through the first calender, the material is heated over an infra-red heating table in order to decrease the viscosity of the resin and facilitate wetting of the fibers. Viscosity values of approximately 10 Pa·s are typical. The heating temperature depends on the resin matrix, the conveying rate of the tape line, and the desired degree of impregnation of the prepreg. Past the heating table, the heated material is fed through several calender rolls, pressing the resin matrix into the fiber bed. These calender rolls may also be heated in order to control the resin viscosity. The distance between the calender (compaction) rolls controls and determines the degree of impregnation of the prepreg. This distance is set by either the roll clearance or by the applied pressure. The fact that the different parameters, such as line speed, temperature, clearance of the calender rolls, and matrix viscosity, mutually influence each other contributes to the complexity and difficulty of the impregnation process. The process temperature is particularly critical, because at each processing step the prepreg resin system will undergo limited reaction, which is promoted by elevated temperatures. To a certain degree, however, some pre-reaction is often desired in order to influence the viscosity of the resin matrix and decrease resin flow during curing and subsequent component manufacturing. These systems are also called controlled-flow systems (B-stage state, see Section 3.2.4 and Chapter 5).

After impregnation, the material is cooled to approx. room temperature on a cooling table and then enters the post-impregnation zone. Depending on the delivery form and material properties, the resin film's carrier paper (process paper) is removed (completely or partially) and replaced by new, silicone-coated paper and/or polyethylene film [3, 4].

The paper or film is necessary to avoid contamination, to make sure the prepreg does not adhere to itself (separation) during storage on cores, and to facilitate subsequent processing steps (hand lay-up or automated processes). Following paper/film lamination, the material is trimmed by rotary knives and can be cut to customer specification (typical widths are 150/300/600 mm). Finally, the material is wound on cardboard cores, hermetically sealed in film bags, and deep-frozen in order to slow down the curing reaction of the resin matrix system.

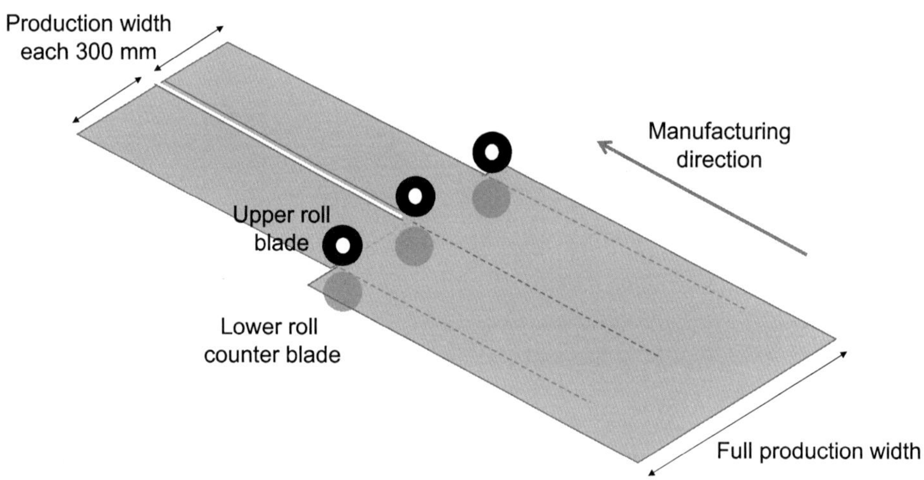

Figure 3.8 Trimming of prepreg to nominal width during processing

The prepreg material's delivery form is determined by the intended subsequent manufacturing process. Automated tape laying materials (Figure 3.9) and materials intended for the manufacture of slit tape (Figure 3.12) are laminated on paper;

hand lay-up materials are often covered with polyethylene film on the other side in order to facilitate trimming (either by cutter or by hand) and also to protect the material from contamination (see Section 4.2). After removal from the prepreg, these materials must not be used for subsequent processing steps and are disposed of. Together with trimming and other production scrap, they are recycled and reused in other products.

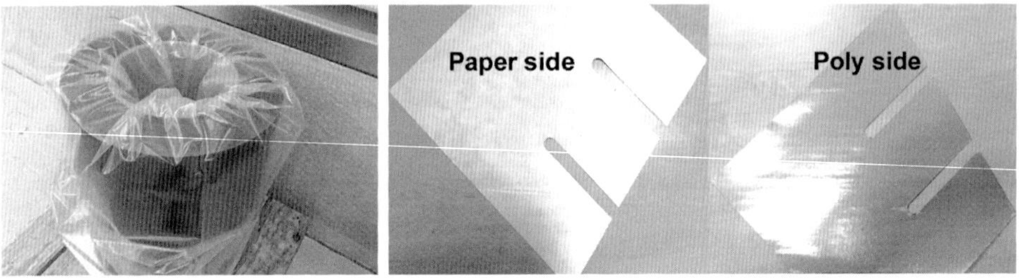

Figure 3.9 Prepreg delivery forms: ATL-prepreg on a core, wrapped in film (left), with separating film and paper (right)

3.2.2.2 Solvent Coating/Solvent Dip Method

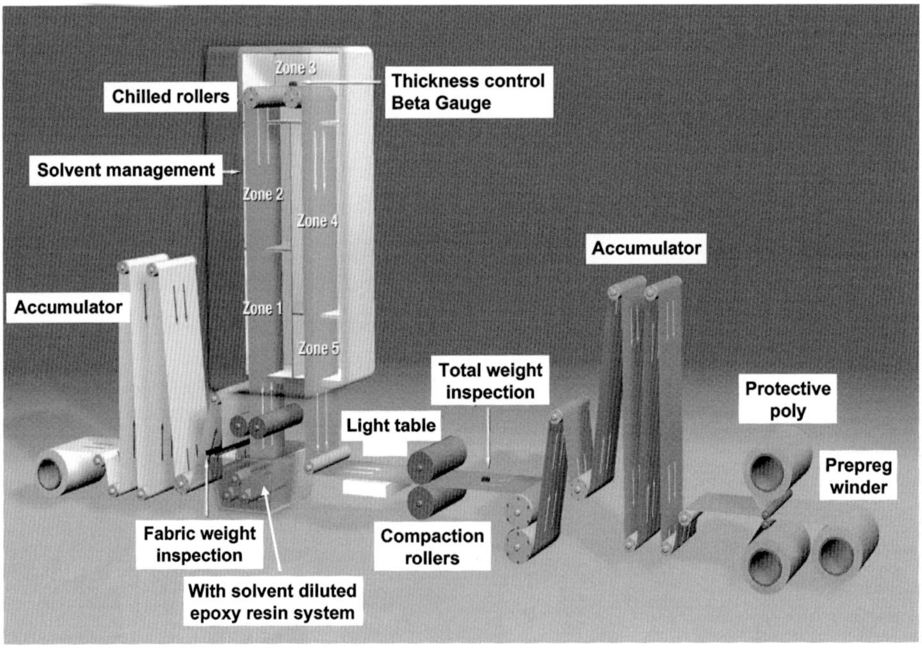

Figure 3.10 Schematic of solvent coating process [Courtesy: © Hexcel Corporation]

The solvent coating process/solvent dip method is used to impregnate fabrics or individual fiber tows (tow prepregs) (Figure 3.10). Here, the creel set used in the hot melt process is replaced by a roll unwinder to feed the semi-finished fabric product into the line. The tension of the fabric is set and controlled (traction control) in order to avoid wrinkle formation and warpage of the reinforcing material. The material is then run through a bath of solvated resin (dip process) where capillary forces ensure the wetting of the material. The viscosity of resin solution is controlled by the bath temperature and the solvent content in the bath. Excess resin on the surface is removed by a doctor blade and collected. Calender rolls are used to control resin content and prepreg thickness. The volatile content in the impregnated material is mostly evaporated by heat and recirculating air in a subsequent drying tower and reclaimed to meet cost and environmental requirements. During this step, the thickness of the impregnated material is tested. Once the volatile content has reached the desired level, the prepreg is cooled by chilled rollers so that resin flow and curing reaction are brought to a halt. After passage through the tower, the material undergoes optical testing (light table) for possible defects before its areal weight is determined. The impregnated material is then deposited on films or silicone-coated carrier paper to protect it from sticking and contamination, wound on cores, and packaged in film bags. Unlike hot melt prepregs, solution coated prepregs are not trimmed; their width is determined by the width of the reinforcing fabric.

The solvent dip process offers certain advantages compared to the hot melt process. For example, here very high-viscosity resin systems or systems with insoluble particles can be processed relatively easily by appropriate adjustment of the solvent content. Also, the fabric wet-out is very good, because the resin almost completely penetrates the fiber filaments and fills the interstices. Prepregs manufactured by solvent coating often exhibit higher tack than comparable hot melt prepregs. Their tack is determined by parameters such as residual solvent content in the prepreg, temperature load in the tower, the resin system itself, and the fiber/fabric type [3, 4].

A disadvantage of the solvent coating process is the fact that it cannot be used for continuous UD-fiber impregnation. A homogeneous UD-fiber bed without fixation by a weft would be destroyed in the solvent bath; the fiber bundles would disintegrate, creating gaps between the fibers. The only solvent coating process available for UD fibers is the use of a so-called drum system, where fiber bundles are wound on quadrangular or octagonal drums and then impregnated using a dipping process. This process is suitable only for small production runs and experimental purposes.

In the past, the hot melt process was not as widely used as the solvent coating process because of its rather difficult process management and limited applications. However, because of changes in environmental regulations and the increasing material and waste disposal costs for organic solvents, hot melt processes have gained increasing acceptance for the manufacture of fabric prepregs since approx. 2005.

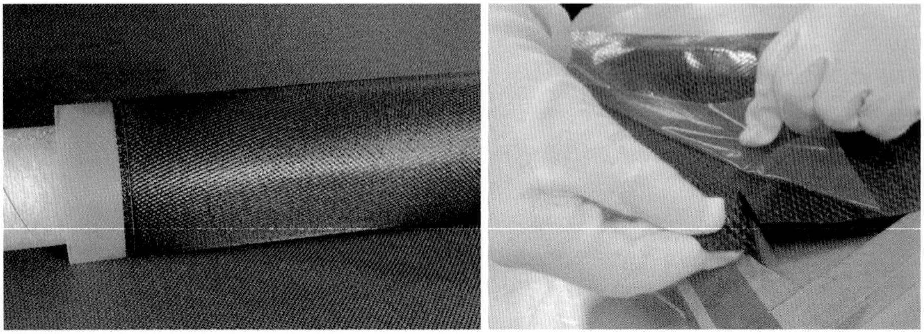

Figure 3.11 Fabric prepreg [Courtesy: © Hexcel Corporation]

3.2.3 Prepreg Variations

Besides the "standard prepreg variations", there are a number of other prepregs, including tow pregs and slit tapes. While the latter have been on the market only since approx. 2006, tow pregs have been available since the beginning of prepreg technology.

3.2.3.1 Slit Tape

Slit tapes (Figure 3.12) are used in automatic fiber placement (AFP) production lines, see Section 4.4. Here, UD prepreg is first produced as so-called parent tape (600 mm or wider) and in an additional cutting process slit into narrow tapes by rotary cutters, applied to a release film, and rewound on spools. Depending on future application, the tapes are provided in widths of typically 3.2 mm (1/8"), 6.4 mm (1/4"), and 12.7 mm (1/2"). Standard spool weights are typically 3 kg, 7 kg, or 11 kg. The advantage of this prepreg variation is its excellent drape over three-dimensional surfaces when using automated processes and its high precision in terms of cutting width tolerances (up to 1/10 mm), independent of the tape width. These narrow tolerances are required in aerospace applications in order to produce high-quality, lightweight components with consistent properties. Because of their easy reproducibility, slit tapes also offer narrow tolerances in terms of FAW and resin content.

Figure 3.12 1/2 ''-slit tape with film release on spool

However, the slitting process with its narrow tolerances results in a considerably higher price per m^2 compared to "normal" UD prepregs. In addition, the prepregs are exposed to additional thermal loads because the tapes are slit at room temperature, without any cooling. This reduces the time for which sufficient tack can be guaranteed. The typical workability of slit tapes on AFP systems is approx. 8 to 10 days.

3.2.3.2 Tow Pregs

Tow pregs are impregnated fiber rovings that are usually wetted in a solution coating process (Section 3.2.2.2) and then wound on a cardboard core. To avoid sticking of the tows, they are wound on a separator film. The rovings are spread (e.g., pneumatically) prior to impregnation in order to achieve a homogeneous and uniform width of the tape and to set the FAW of the finished tow preg. This type of prepreg is particularly suitable for filament winding processes and often manufactured in situ and immediately processed into final components. Its advantage compared to standard prepregs and slit tapes is its considerably lower price and its flexibility regarding suitable resin systems. The only requirement on the resin system is adequately low viscosity (if necessary by solvent addition) to achieve sufficient impregnation of the fiber tows. In contrast to standard prepregs and slit tapes, the width of the tow preg cannot be controlled within narrow tolerances, but rather varies with roving width [5, 6].

3.2.4 Effect of Manufacturing Parameters

Prepreg properties are affected by the selected processing parameters and processes. The following will describe and discuss some of these properties and the parameters influencing them.

3.2.4.1 Resin Content

The resin content of a prepreg is relatively easy to set and control. Together with the FAW it defines the cured ply thickness (CPT) of a prepreg ply after curing. The resin content of a prepreg is given in wt.% relative to the FAW (Section 3.2.2.1) and provides the total weight of the prepreg when added to the FAW.

Depending on resin system and degree of impregnation, the resin content determines the tack and resin flow (bleeding of excess resin) of the prepreg. While the prepreg systems of the 1st and 2nd generation often had a higher resin content and therefore exhibited higher resin flow during curing, the current 3rd-generation prepreg systems are manufactured as so-called zero-bleed prepregs in order to optimize component weight. Here, the resin content is precisely set to match the requirements for fiber volume content and ply thickness of the final component.

Table 3.1 and Figure 3.13 exemplify how the physical properties of a prepreg change when resin content is varied from 35 wt.% at constant areal weight (e.g., 134 g/m²). Significant changes are seen in cured ply thickness (CPT) and fiber volume content (FVC).

Table 3.1 Physical Prepreg Properties [Courtesy: © Hexcel Corporation]

	Unit	Physical prepreg properties												
Prepreg areal weight	g/m²	197	199	200	202	203	205	206	208	209	211	213	214	216
Resin content	%	32.0	32.5	33	33.5	34	34.5	35	35.5	36	36.5	37	37.5	38
Resin areal weight	g/m²	63	65	66	68	69	71	72	74	75	77	79	80	82
Fiber areal weight	g/m²	134	134	134	134	134	134	134	134	134	134	134	134	134
Resin volume content	%	39.30	39.85	40.39	40.94	41.48	42.02	42.56	43.09	43.63	44.16	44.69	45.22	45.75
Fiber volume content	%	60.70	60.15	59.61	59.06	58.52	57.98	57.44	56.91	56.37	55.84	55.31	54.78	54.25
Cured ply thickness (CPT)	mm	0.123	0.124	0.126	0.127	0.128	0.129	0.13	0.132	0.133	0.134	0.135	0.137	0.138
Composite density	g/cm³	1.60	1.60	1.59	1.59	1.59	1.58	1.58	1.58	1.58	1.57	1.57	1.57	1.57

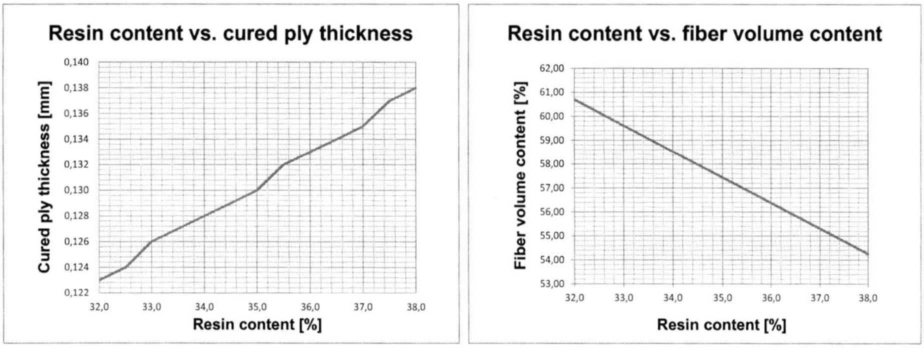

Figure 3.13 Changes in CPT and FVC with prepreg resin content

3.2.4.2 Level of Impregnation

Another important parameter is the level of impregnation of the prepreg. It indicates to what the degree the resin matrix has wetted the fiber material during the manufacturing process.

The level of impregnation and its adjustability are primarily dependent on fiber and resin areal weight as well as on the impregnation method. The adequate impregnation of the reinforcing fiber influences not only the resin distribution during curing, but also its later processibility, e. g., by automated processes.

Solvent coating processes generally provide fully impregnated prepreg materials. For fabric prepregs, this is desirable in order to eliminate air pockets and the formation of porosities in the final component during subsequent processing, because it is typically not possible to remove air trapped in fabric interstices by in situ post-impregnation during curing. However, this challenging material behavior is also affected by other parameters such as component geometry and size, and process management during curing.

Hot melt processes allow adjustment of the degree of impregnation within certain limits, because here resin film composition and weight, FAW, and impregnation temperature and duration can be varied. In contrast to fabric prepregs, many UD prepregs are not fully impregnated, but contain a thin interior area of dry filaments that allows for resin flow and thus for air removal. The use of fully impregnated UD prepregs may lead to the formation of porosities in the composite part, in particular in large-area components.

Figure 3.14 shows some parameters and constraints influencing the level of impregnation:

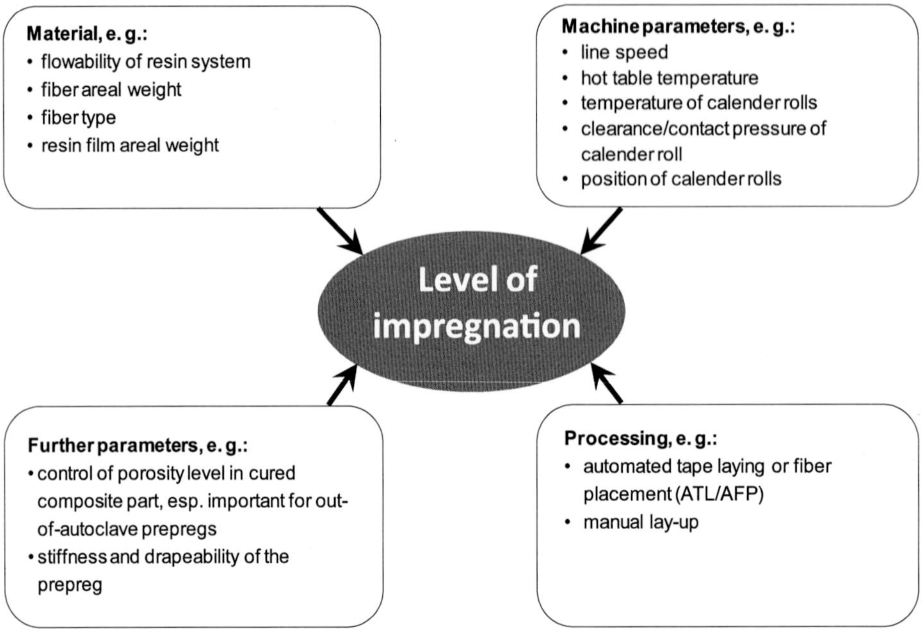

Figure 3.14 Examples of parameters and constraints influencing the level of impregnation in hot melt prepregs

It is not possible to record the level of impregnation as an online parameter during the prepreg manufacturing process. Therefore, water pickup and other methods are often used to determine the level of impregnation (see Chapter 7) and to control and manage the impregnation process.

Figure 3.15 shows the cut edges of prepregs with different degrees of impregnation (approx. 6 % and 1 % water pickup). These so-called "false-color images" assign a color value to each gray value in a black/white photograph. This color representation allows for an automated determination of the level of impregnation by photo analysis as well as establishing a correlation with the water pickup values, facilitating a quick and simple inspection during processing.

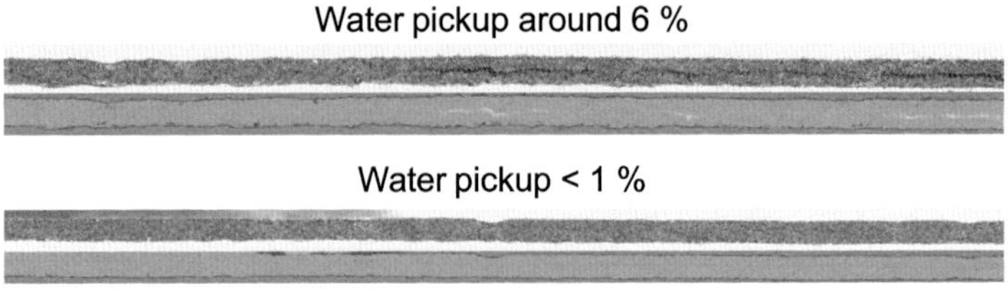

Figure 3.15 Cut edges of prepregs with different levels of impregnation

The level of impregnation affects other properties, such as drape and tack of the prepreg. At higher degrees of impregnation, the uncured material is stiffer and therefore decreases the drape of the material. At lower levels of impregnation, the material feels softer and it is easier to drape, e.g., over radii. Second-generation prepregs are often used with slight excess resin and resin flow is intended to achieve a specific fiber volume content in the cured component so that a higher level of impregnation can be set. In 3rd-generation prepregs, the percentage of resin content is already set precisely to achieve the fiber volume content in the cured component (58 to 60%). Here, a lower level of impregnation is specifically chosen to ensure sufficient resin on the fiber surface and thus sufficient tack for subsequent processing (Section 3.2.4.3 “Tack”).

Determining the exact product composition is an iterative process, because the different parameters are interdependent. Ultimately, factors such as the manufacturing process, the additives used for the manufacture of the final component, as well as component geometry and size, have decisive influence on the desired degree of impregnation.

3.2.4.3 Tack (Tackiness)

The term “tack” describes the ability of different material layers to adhere to each other or to a mold surface. This property is required for the manufacture of composite materials in order to prevent slipping of individual plies while facilitating peeling off for necessary corrections. With high tack, out of place plies can no longer be peeled without destroying them; with low tack, the plies may slide.

There is no standardized scale for the tackiness of a material because of the wide variety of materials commercially available, nor are there universal testing methods. Because tack is an important property for other materials and semi-finished products, such as coatings, a variety of methods and testing procedures, each limited to a very specific application, have been established.

As seen in Figure 3.16, the tack of thermosetting prepregs is determined by a variety of different parameters.

It is obvious that, in contrast to the level of impregnation, the tack of a material cannot easily be set or controlled, but is a property influenced by multiple parameters. In addition, these parameters influence each other, e.g., base material and manufacturing process and/or storage stability, which in turn affect processing.

Raw materials, such as resin base and modifier content in the prepreg resin system, may have the more significant influence on the later tack of a prepreg. The selection of raw materials is almost exclusively determined by the requirements on the properties to be achieved in the cured state.

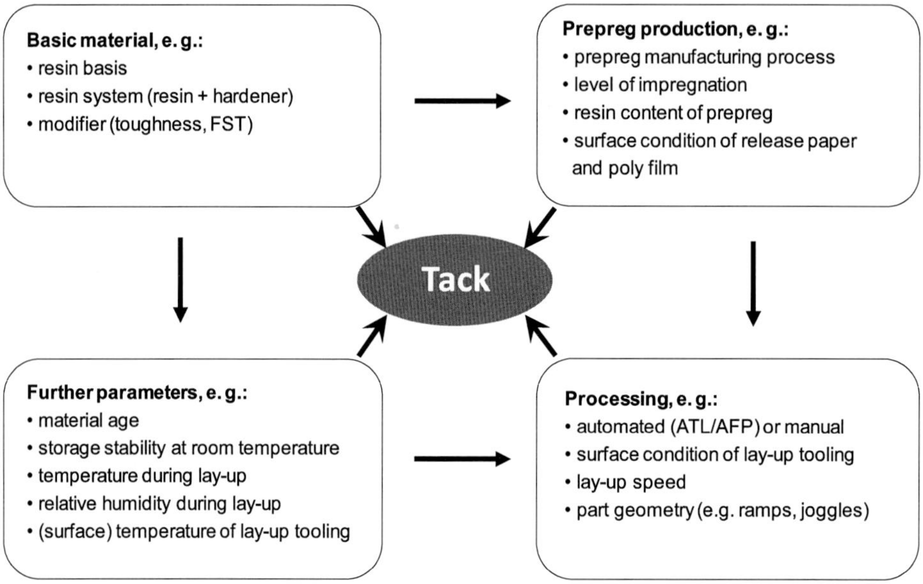

Figure 3.16 Examples of parameters affecting tack

The process used to manufacture the prepreg also influences its tack, because manufacturing method and subsequent application result in numerous influencing parameters that affect the further processing of prepreg materials. At this point, it becomes apparent whether the material's tack is suitable for the chosen processing method. If that is not the case, adjustments can only be made in terms of the processing method, because the material properties have to meet specifications and must not be changed. Adjustments in the prepreg manufacturing process would require an expensive re-qualification of the material. In particular in the aircraft industry, prepregs are designed to meet specific characteristics, such as glass transition temperature, toughness, and extremely low resin flow. These characteristics determine the appearance and processibility of the material. The processibility and tackiness of the prepreg material are also affected by mold surface and lay-up rate.

Also to be considered are the influencing effects of material aging and ambient conditions during processing (relative humidity and temperature). In particular the age of a prepreg material has decisive influence on its tackiness [7].

The processor has to consider a wide variety of parameters that affect processing as well as the characteristics of the cured material.

3.3 References

[1] A. Glawe, K. Timmermanns, R. Reuscher. *Beschichtungs- und Verarbeitungstechnologien zur Herstellung von Prepreg-Produkten*. Technische Textilien. 2008, 4, 174–175.

[2] John W. Putnam, Brian S. Hyes, James C. Seferis. *Prepreg Process-Structure-Property Analysis and Scale-Up for Manufacturing and Performance*. Journal of Advanced Composites. 1996, 47–57.

[3] James C. Seferis, Roman W. Hillermeier, Frédéric U. Buehler. *Prepreging and Autoclaving of Thermoset Composites – Comprehensive Composite Materials*. (ISBN: 0-08 0429939). Volume 2 (ISBN: 0-080437206), 701–736.

[4] Company brochure: *Prepreg Technologies*. Hexcel Composites. 2012.

[5] High Performance Composites: *From Art to Science: A Prepreg Overview*. 2000, May/June. www.hpcomposites.com

[6] K. Kawabe, H. Sasayama, S. Tomoda. *New Carbon Fiber Tow-Spread Technology and Applications to Advanced Composite Materials*. SAMPE Journal. 2009, 45, 6–17.

[7] R.J. Crossley, P.J. Schubel, N.A. Warrior. *Experimental Determination and Control of Prepreg Tack for Automated Manufacture*. Plastic, Rubber and Composites. 2011, 40, 363–368.

4 Prepregs: Processing Technology

Hauke Lengsfeld and Javier Lacalle

4.1 Introduction

In this chapter, we will describe the different technologies used to process prepregs and to transform them into prepreg components. We will introduce both manual and automated deposition methods as well as methods to cut and form prepreg materials.

One of the advantages of fiber reinforced materials is the fact that the fiber reinforcement can be strategically placed in the component to optimize the relationship between mechanical properties and weight.

Manufacturing methods using prepreg are particularly efficient because they achieve highest quality and most accurate fiber placement in the component. This also ensures optimum fiber volume content in the component, because the ratio of resin to fiber has already been coordinated in the prepreg. The layer structure may be deposited by hand or by automated processes.

In general, the processes used to manufacture composites are laminating and deposition processes that use flat, semi-finished products (e. g., prepregs) and deposit them in a specific sequence and in a defined orientation and shape on a mold or tool.

Today, there is a wide variety of processes available. There are two different approaches to classify these processes. First, considering the pressure applied during forming and curing leads to the classification of the most common processes as shown in Figure 4.1:

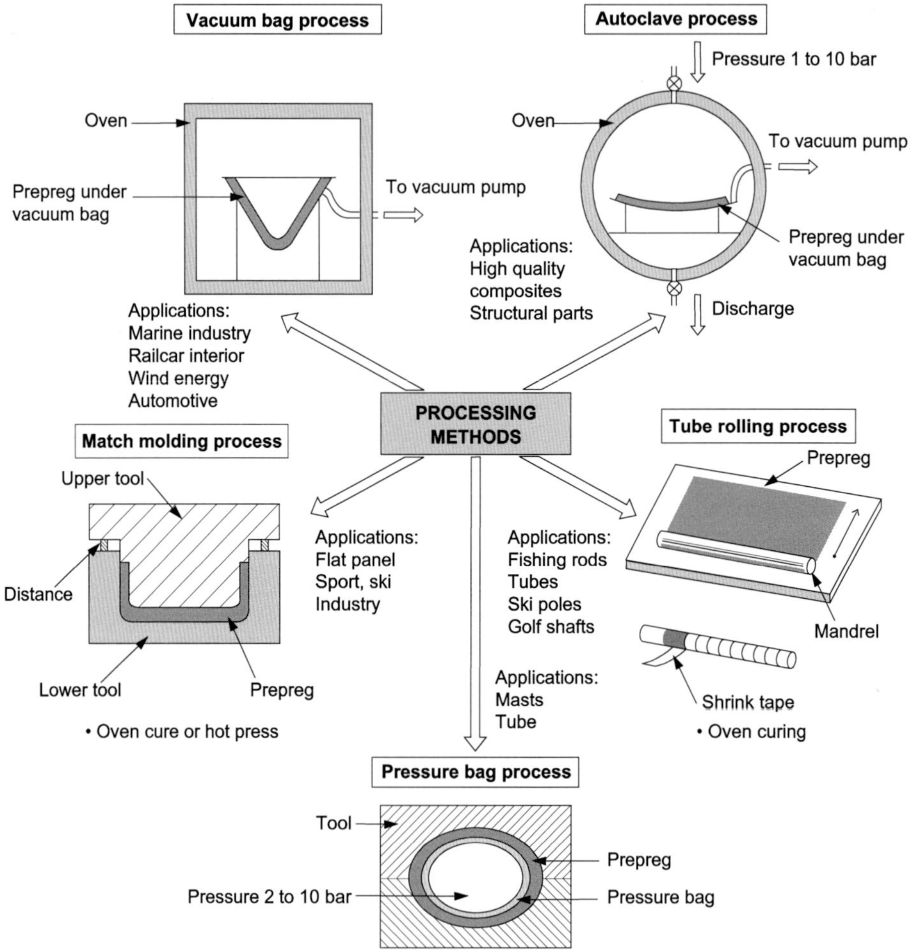

Figure 4.1 Curing and forming processes for prepregs [Courtesy: © Hexcel Corporation]

An alternative perspective considers the deposition and processing technologies that are used to process prepregs (Figure 4.2). This approach also includes the subsequent process of manufacturing a fiber reinforced composite part and provides a differentiated classification in the various technologies. While the pressure-oriented approach (see Figure 4.1) often includes the curing process, it is deliberately separated from the deposition and processing technologies. Curing is considered a separate step and thus includes the pressure processes shown in Figure 4.1.

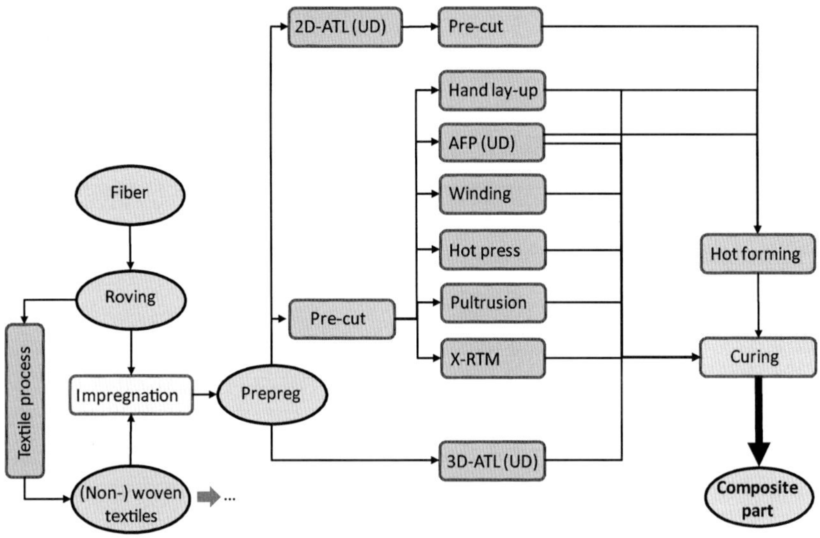

Figure 4.2 Deposition and processing technologies for prepregs

Figure 4.2 shows a variety of processing technologies for prepreg materials. Not every technology is equally suitable for all prepreg types. For example, resin impregnated fabrics cannot normally be deposited using ATL or AFP processing technologies. In contrast, UD prepregs can be processed into mats or components using any of the available process technologies.

The choice of different ply configurations during deposition allows for the manufacture of components with varying thickness and structure. In general, each separate layer (ply) may exhibit a different contour and fiber orientation (Figure 4.3).

In the following, we will discuss in more detail the processing steps shown in Figure 4.2 from the time at which the prepreg is provided.

Figure 4.3 Laminate consisting of net shape cut layers (plies) (left), and three-dimensional ATL deposition (right)

4.2 Cutting of Prepregs

Prepregs are commercially available on rolls (see Chapters 2 and 3), and the roll widths may vary widely depending on material and customer requirements. UD tapes are typically provided in widths of 150, 300, 600, and 1200 mm, while fabric prepregs come in widths of 1000 to 1270 mm.

Prior to manufacturing of the final component, the material has to be cut to match the desired geometry (see Figure 4.2). Here, we distinguish between two major methods: manual cut or automated cutting systems. Cut parts are often "kitted" so that several parts that belong together can be packed in one box. The individual cut parts are labeled so that they can be identified and processed later (either via computerized systems or with the help of drawings). Because prepregs have only limited out times (storage or processing at room temperature), the kits can be frozen and processed at a later time. Cutting of resin imbedded carbon fiber layers or fabrics is relatively easy because of their low shear strength and buckling resistance, which is a distinctive advantage compared to other fibers. However, the abrasive character of carbon fibers causes wear of the cutting tools, in particular non-hardened tools.

4.2.1 Manual Cutting

For manual cuts, a steel knife (cutter, Stanley knife) is traced along a straight cutter bar or along a metal template for contoured cuts. Manual cutting is typically used for relatively small geometric dimensions or when the number of cut pieces does not justify automation. However, efficiency and accuracy of this method depend on the dimensional accuracy of the template as well as on the experience of the personnel.

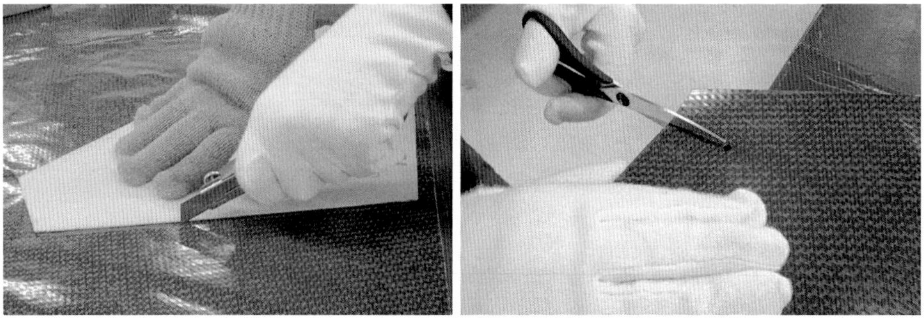

Figure 4.4 Manual cut by cutter or scissors

4.2.2 Automated Cutting

In the early days of automated prepreg cutting, both ultrasonic and water jet cutting systems were employed. Today, the latter are no longer in use because the moisture introduced in the prepreg by the water jet often caused the formation of porosity during curing of the composite. In modern production lines, automated cutting is typically performed by ultrasonic cutters. They are computer-controlled machines designed for cutting of textiles and other pliable materials. The first CNC-cutters were developed in the 1970s for textile applications, and subsequently optimized for cutting of dry glass fiber and carbon fiber reinforced materials and prepregs.

Automated Cutting Systems

In principle, a cutter consists of a plane cutting table of variable size (depending on application) and a traversing X-Y-system above it (Cartesian coordinate system) that carries the cutting head (Figure 4.5). The cutting path is generated by the superimposed motion of both axes. The cutting head may also contain a labeling system to label each cut and/or a camera to control the correct position of the material.

The material is cut by a knife (Figure 4.6) that is moved into the appropriate working position along the z-axis. A variety of different knives is used, e.g., cutting wheels (both powered and non-powered), oscillating knives, drag knives, and punch knives. Some manufacturers ultrasonically actuate their knives to prevent sticking of the blade to the pre-impregnated material and to simultaneously assist cutting of the abrasive fibers. The finished cuts (Figure 4.7) are removed from the cutting table either manually or by a vacuum gripper and supplied to the next processing step.

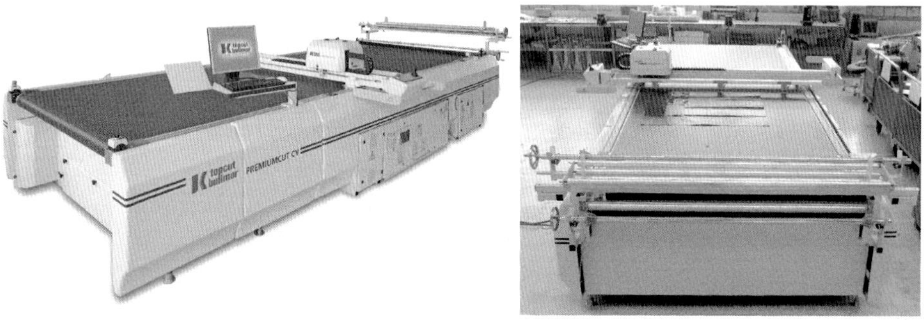

Figure 4.5 Cutter systems [Courtesy: assyst Bullmer GmbH, Gunnar/Weissenberger AG]

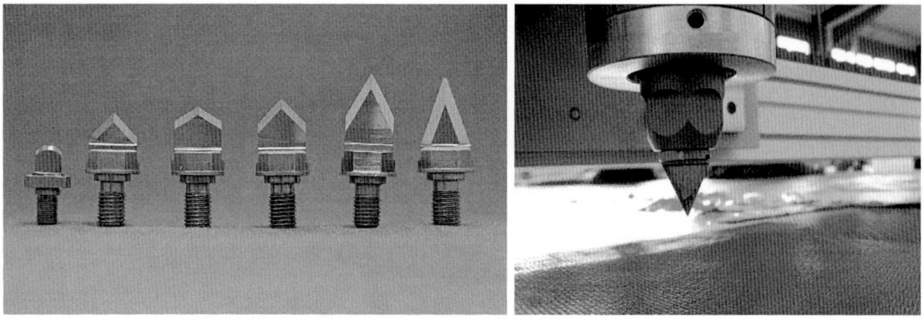

Figure 4.6 Cutter knives (drag knives) [Courtesy: MUK GmbH Elmshorn (left), assyst Bullmer GmbH (right)]

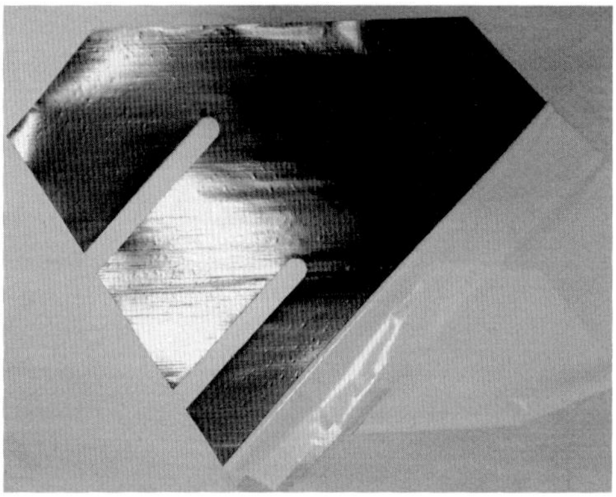

Figure 4.7 Finished cut

The type of knife chosen for a specific application depends on the material to be cut and the cut geometry. Cutters with a steep blade geometry are used in order to produce a sack or bag blank (see Figure 4.8) without cutting into the remaining material or at least minimizing the incision (Figure 4.8, right).

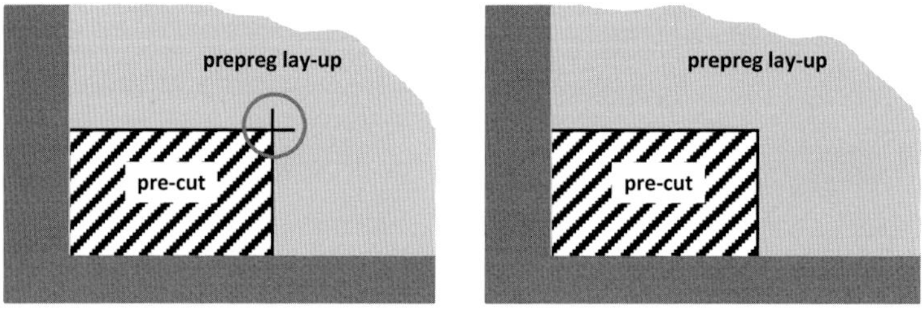

Figure 4.8 Cutting bag blanks using standard knife (left) and special knife (right)

The cutting table is typically covered with a cutting surface made of felt, plastic, or bristle blocks that allow the cutting tool to slightly penetrate the cutting surface, thus facilitating a clean cut (see Figure 4.9). Depending on model, the cutting surface can be fixed to the table top or it can take the form of a conveying surface that simultaneously transports the material to be cut. In both cases, the table will also be equipped with a vacuum suction system. For conveyor cutting systems, the cutting table is typically extended in the conveyed direction beyond the working area of the cutting head to facilitate the removal of the finished cuts (Figure 4.5). A cleaning mechanism for the conveyor belt is typically integrated in the cutting table. On the side of the cutting table at which the material is fed into the work area of the cutting head, supply systems for automated material feeding (either from rolls or from sheets) can be integrated.

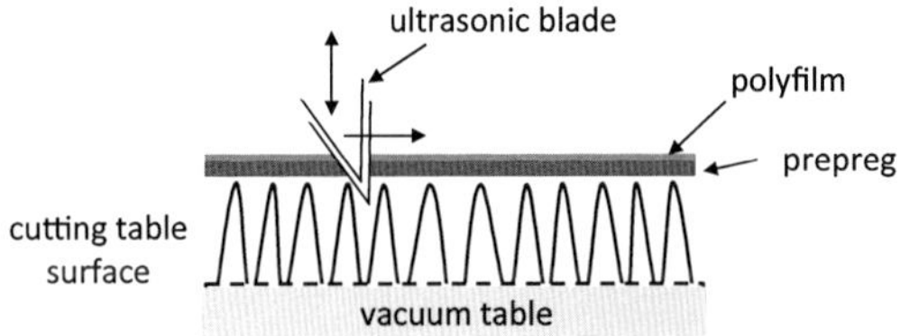

Figure 4.9 Cutting surface (schematic)

The cutting head is positioned on the y-axis of the drive, serving as a tool carrier, in general equipped with a tool changing system, simultaneously holding several blades in quick-change mounts. The tool mounts can be equipped with tangential controls to allow active blade control for movement tangential to the cutting direction. Cutting blades have proven to be the best compromise when cutting textile materials. When cutting fiber reinforced semi-finished products, and prepregs in particular, the use of water jet and laser systems should be avoided because of the undesired introduction of moisture and thermal energy into the material.

Nesting

The use of automated cutting systems offers significant advantages compared to manual cutting operations both in terms of dimensional accuracy of the cuts (higher precision) and maximum material utilization by optimal nesting of cuts on a given material width. This nesting (Figure 4.10) is often supported by the CNC software of the cutter systems. Automated systems typically cut single-ply prepregs (UD tapes and fabrics), multi-ply prepreg stacks, and dry fiber reinforced materials (single- and multi-ply).

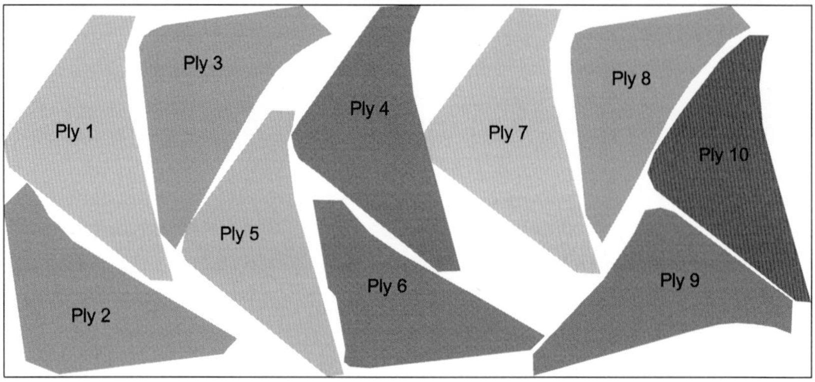

Figure 4.10 Nesting pattern on a fabric layer for an NC cutter

The NC data required for the cutter are derived from a composite part design (CPD) model. In Step 1 (Figure 4.11) a "solid" of the finished component is designed using CAD software, such as CATIA, without defining the layer geometries or the stacking. This step defines the desired component geometry. In Step 2 (Figure 4.11), single-ply thickness and boundaries are defined by slicing the solid by planes separated by single-ply thickness. This step defines the single-ply geometries and contours.

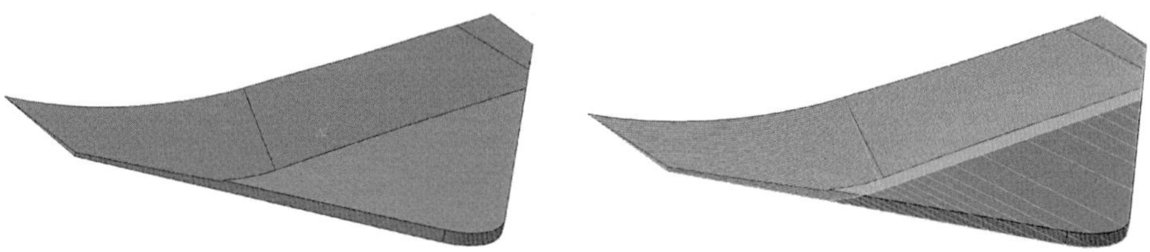

Figure 4.11 Step 1: solid; Step 2: slicing [Courtesy: CTC GmbH]

Next, the desired ply orientation is assigned to each ply geometry: 0°, 45°, 90°, or 135°.

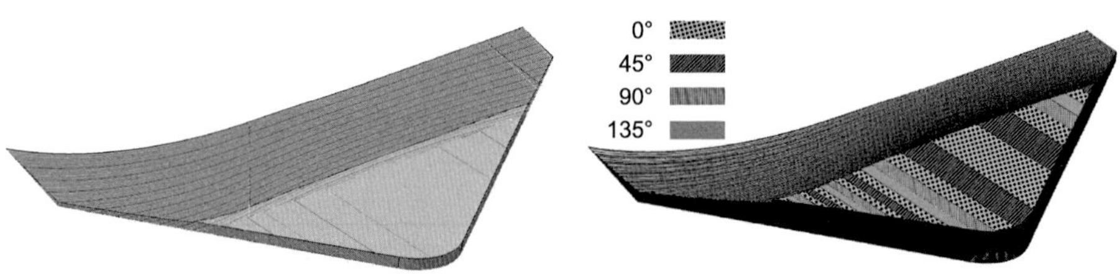

Figure 4.12 Ply orientation and exploded plies [Courtesy: CTC GmbH]

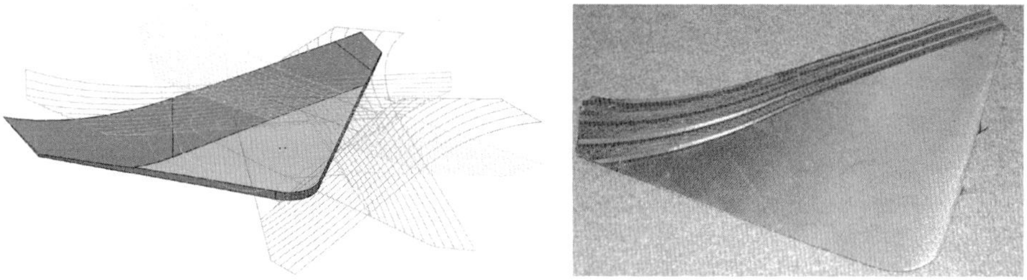

Figure 4.13 Developed view of the ply cuts and finished cut [Courtesy: CTC GmbH]

The plies are represented as exploded plies (Figure 4.12). Each colored surface represents one ply. Subsequently, the ply geometries are transferred to a plane in order to fabricate the cut pieces and the plies are aligned depending on fiber orientation. Finally, the geometries are exported into a data format suitable for NC cutters or tape layers.

4.3 Hand Lay-up of Prepregs

Hand lay-up is the simplest way to use prepregs and produce a composite component (see also Figure 4.15). Here, the prepreg cuts (UD tapes or fabrics) are stacked (laminated) on a tool following a ply book that exactly specifies the number of plies and their angles. Then they are pressed together to achieve good bonding and to prevent air entrapment.

Prepreg materials are delivered on rolls. The prepreg is unwound together with the carrier paper and manually cut to exact shape using a guide and cutter. This process step usually results in scrap. The carrier paper is removed and the cut prepreg is manually (sometimes with the help of auxiliary tools) positioned on the component to be built. The initial ply adhesion should always allow for the non-destructive removal and correction of an incorrectly positioned layer. However, this is always dependent on the tack of the utilized prepreg (see Section 3.2.3). Layer deposition should always proceed slowly from one side, uniformly rolling out the layer to avoid air entrapment. Rolling can be carried out by hand; however, the use of rollers results in more uniform consolidation. Air entrapped between the layers can be pushed out with the help of a plastic putty knife or by carefully tapping the prepreg with a needle. Another option to remove air and simultaneously consolidate the prepreg structure is the application of vacuum. Here, the prepreg structure is kept under vacuum for approx. 10 to 30 min at a time at room temperature. This processing sequence for the deposition of a tape is schematically shown in Figure 4.14.

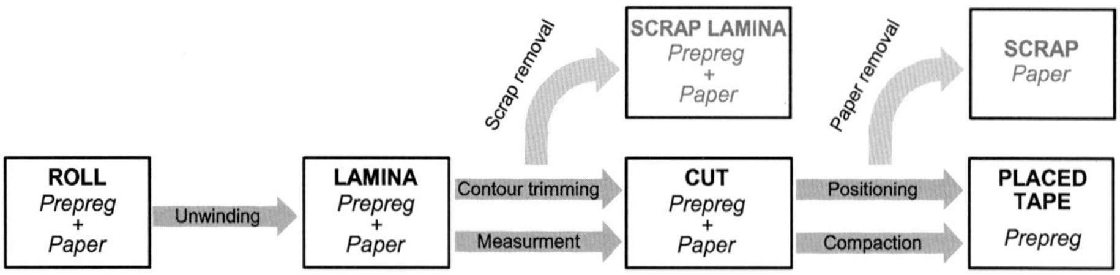

Figure 4.14 Schematic sequence for manual tape deposition

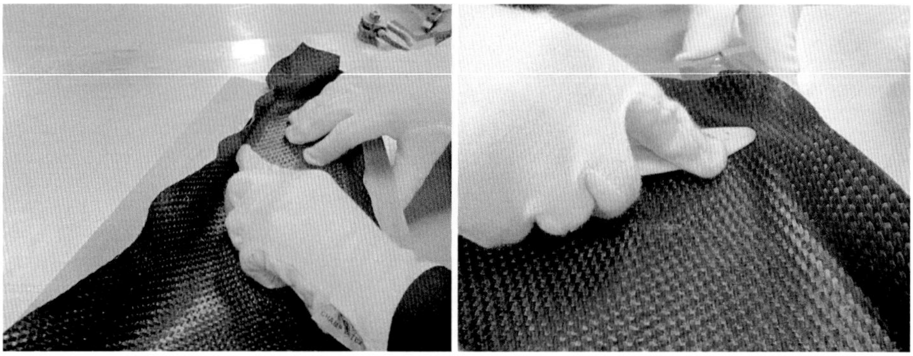

Figure 4.15 Hand lay-up: process steps

In addition, the prepreg can be heated to 50 to 60 °C in order to lowering the resin viscosity and thus support the process.

Hand lay-up is typically used for smaller components and/or component parts with highly sophisticated geometries/contours that cannot be economically manufactured using automated processes. Often, such components or component parts are manufactured using fabric prepregs because they offer better drape than unidirectional tapes. Currently, there are no automated industrial processes for prepreg fabric draping; however, there are several processes using robot technology for the automated handling of dry fabric cuts.

In general, only relatively thin (approx. 3 to 8 mm) flat, multi-ply prepreg structures can be (hot-)formed to create three-dimensional contours, because multiaxial prepreg structures are typically inherently rather stable. It is difficult to form prepreg structures with tight curvatures and/or complex ply compositions smoothly and wrinkle-free, even when applying elevated temperatures to decrease resin viscosity and improve interlaminar slip. The result often is disrupted fiber orientation/alignment.

During hand lay-up, the cuts are typically aligned using templates, leading edges, or markings on the edge of the mold to achieve sufficient deposition accuracy (positional tolerance) of the individual plies and to avoid incorrect deposition.

Direct labeling of the finished cuts during cutting (whether manual or by cutter) prevents the interchange of cuttings during lamination. In addition, nowadays laser positioning systems are used that project the exact position and form of individual cuts on the lay-up area. These systems also prevent turning of cuttings and lay-up in the wrong orientation. The disadvantage of using templates becomes apparent when many different-sized cuts require the same high number of templates. With larger cuts, templates tend to become rather unwieldy. Another problem with hand lay-up of prepreg cuts is the sheer number of prepreg cuts necessary for the manufacture of complex components, which can reach several hundreds. This results in an additional logistics problem, especially when the cuts cannot be stored at room temperature but have to be kitted and frozen for storage.

Several parameters, including size, production rate, and the required accuracy and repeatability of the prepreg lay-up (form and positional tolerances) determine whether a component will be manufactured by hand lay-up or by an automated process. On the one hand, hand lay-up processes are time and personnel intensive; on the other hand, the high capital and operating costs of automated processes have to be considered.

In the past (until approx. 2004), large components, such as the vertical tail plane of the Airbus A320 and A330 series, were manufactured by hand lay-up of unidirectional tapes. For large components, the handling, exact positioning, and laying of long prepreg cuts without entrapping air is extremely difficult, and defects in the composite part are therefore hard to avoid. Therefore, the prepreg structure of such large components is typically manufactured using automated systems.

4.4 Automated Laying Systems: Automated Tape Laying (ATL) and Automated Fiber Placement (AFP)

4.4.1 Introduction

Today, the automated laying of pre-impregnated fiber materials is a key technology for the manufacture of large composite components in the aerospace industry. For a number of years, automated tape laying has been used in conjunction with other technologies, such as hot-forming, to manufacture vertical tail panels, wing structures, stringers, spars, etc. The fast growing use of composite components both in aerospace and automotive applications, together with the increasing complexity of these components, has triggered the continuous growth in use and research of automated and highly efficient laying technologies. Figure 4.16 and Figure 4.17 show the geometric complexity as well as the size of composite parts.

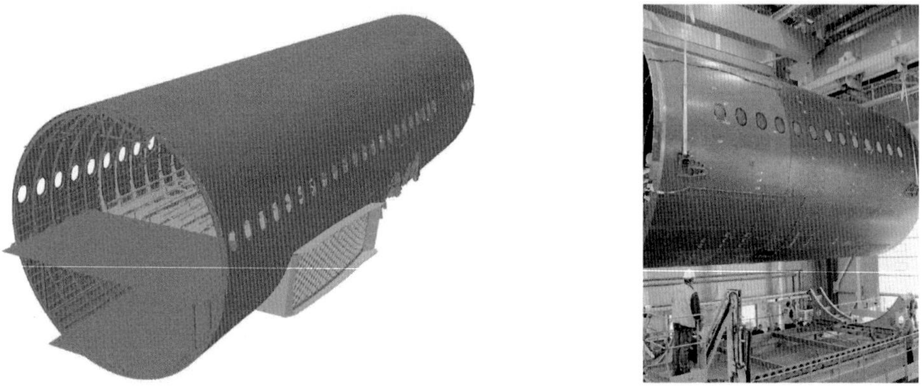

Figure 4.16 Carbon fiber reinforced composite fuselage structures [Courtesy: Airbus Operations GmbH]

The use of automated tape laying (ATL) and fiber placement (AFP) systems for the processing of UD tapes offers significant quality and productivity advantages compared to hand lay-up processes:

- Ply positioning and repeatability
- Deposition rate
- Uniform and void-free pre-compaction, etc.

Automatic lay-up by ATL und AFP is a widely-used standard process for the production of large parts in the aviation industry, one of the principal customers of prepregs. Both lay-up technologies deposit resin impregnated continuous fibers on the surface of a tool or mold. However, both processes have specific characteristics that make them ideally suited for various applications. Criteria for selecting a specific process include component geometry and manufacturing requirements such as the following:

- Component curvature
- Height and slope of the ramps
- Material drape
- Positive or negative (male or female) tool, etc.

ATL technology is typically used for geometrically simple contours, such as wing skins and flaps, panels for vertical and horizontal tail planes, flat laminates for hot-formed components, etc. AFP technologies are preferred over ATL for geometries with sections of double curvature and variable laminate thickness, such as spars and fuselage panels, etc., owing to their superior flexibility.

A wide variety of component types can be laid up by automated processes, a fact that is underscored by the continuous development of new and advanced lay-up technologies. The standard component types for ATL and AFP technologies include, but are not limited to:

- *Panels and skins:* 2-D panels and laminates, as well as fuselage panels with double curvature, wings, shells for horizontal and vertical tail planes, etc.
- *Closed components:* fuselage segments, tail cones, etc.
- Spars, ribs, stringers, etc.

It is obvious that the manufacturing processes have to meet the requirements of many different components. Therefore, the major challenge is the development and implementation of systems and lay-up technologies that are able to reproduce each of the different process and geometry demands, ranging from monolithic components to sandwich structures.

The high mechanical properties required for aerospace applications demand equally high accuracy and repeatability of the lay-up process. The total volumetric positioning and contour accuracy typically ranges below ± 2.5 mm over the entire lay-up area, which can reach a length of 30 m and a width of 8 m as in the case of the fuselage segment or wing skin seen Figure 4.17.

Compared to hand lay-up, automated lay-up technologies allow for significantly closer tolerances in terms of positioning, gaps, and overlap of parallel courses, resulting in higher mechanical properties. At the same time, automated lay-up technologies have to meet general expectations in terms of productivity for each specific component type and size.

Currently, a wide variety of prepreg manufacturing systems and configurations are available. Crucial for adequate productivity and cost effectiveness is the adaptation of manufacturing system and lay-up technology to the specific components to be produced. The productivity of a manufacturing system largely depends on the geometry and the layer structure of the component. A sub-optimal configuration may have serious negative effects on productivity.

4.4.1.1 Goal of Automated Lay-up Processes

Although the manual prepreg lay-up process described in Section 4.3 is simple and cost effective in terms of auxiliary materials and equipment, it also has significant disadvantages in many industrial processes:

- The *productivity* of the process is very limited and decreases even further when manufacturing complex components.
- Both *accuracy* of the laminated structure and *quality* of the manufactured components are relatively low and depend on the experience of personnel. Processes that require high accuracy therefore also require highly trained personnel and

suitable testing methods to ensure optimal quality. Depending on the quality requirements, a manual process may result in high rates of defective/scrap parts.

- *Component size* is very limited and the required quality standard may only allow for the production of small components.
- Stable processes require *repeatable* and *retraceable* procedures, none of which are offered by manual processes.

It is the goal of automated processes to avoid or minimize the disadvantages of manual lay-up methods mentioned earlier so that the positive characteristics of these methods can still be utilized for a wide spectrum of parts.

Automated processes rely on a variety of approximations and lay-up principles. In particular the manufacturing methods employed in the aerospace industry can be distinguished in two major technologies:

- Automated tape laying (ATL, Figure 4.17 left) and
- Automated fiber placement (AFP, Figure 4.17 right)

In the following sections we will introduce both technologies, describe their characteristics and specifications, and evaluate their utilization in a variety of applications. We will limit our description to the processes specific for thermoset materials because they represent the established methods in the aerospace industry. In general, it is possible to design ATL and AFP technologies to accommodate the processing of other materials with different fiber type and matrix systems (e.g., thermoplastics, materials with added binders, or textiles).

Figure 4.17 Automated lay-up processes: ATL (left), AFP (right) [Courtesy: M.Torres Diseños Industriales S. A. U.]

4.4.1.2 Principle Process Steps of Automated Lay-up Technologies

The major process steps of automated lay-up technologies, from geometry development to final lay-up of a component and the subsequent processing and handling post lay-up, will be described in this section. These general steps are encountered in all major lay-up technologies (e.g., ATL und AFP). Each method has process-specific properties that will be discussed in more detail in the following chapters.

Geometry Definition, Ply Book Compilation, CAD Model, and Lay-up Tools

In contrast to the manufacturing of metal components, the materials used for the production of fiber reinforced components do not exhibit the mechanical and physical properties of the final composite material. Rather, the fiber reinforced component reaches its final properties and shape after the curing stage. This leads to a change in material properties and even a distortion of the component geometry compared to the geometry prior to curing. These effects (spring-in or spring-out) have to be considered during geometry development, and the lay-up geometry has to be adapted in order for the cured and demolded component to exhibit the desired final geometry.

With these aspects in mind, components are manufactured using prepreg technology by sequentially laying down individual prepreg layers with different contours and fiber orientation.

The properties of each prepreg layer are uniquely defined in a ply book for the lay-up process. The ply book is typically provided as a CAD model (Figure 4.18) and can be used to program automaticlay-up processes. It contains the following information:

- Ply number and identification
- Ply contour
- Material
- Ply surface and lay-up side and/or drape orientation
- Definition of fiber orientation and reference coordinate system

Typically, commercially available software packages are used to determine the component geometry and surfaces as well as to compile the ply book.

Offline Programming, Process Simulation, and Machine Programming

The goal of the lay-up process is to make sure that the surfaces and contours defined in the ply book are completely covered with prepregs with a specific fiber orientation and according to specific design rules. The lay-up should also minimize scrap, be adjusted to the theoretical contour (depending on the properties of the lay-up method), and be optimized in terms of productivity and quality.

Software for offline programming is used to achieve this goal. These graphics-based software products allow for user-friendly generation of courses and of complex lay-up strategies. Besides commercially available programs, there are also programs directly developed by the machine manufacturers and provided specifically to match their systems (Figure 4.18).

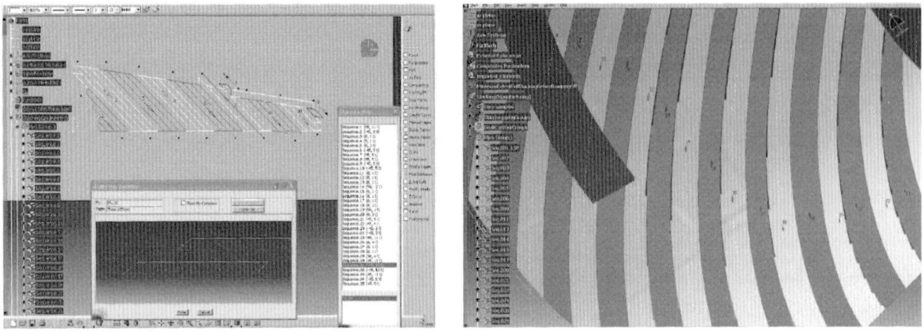

Figure 4.18 Example of ply book structure with ply sequence, contour, and orientation for ATL (left) and AFP (right) [Courtesy: M.Torres Diseños Industriales S. A. U.]

The programming stage takes into account numerous parameters and strategies that are important for the subsequent lay-up process. Some of these options include:

- Initial start of lay-up and staggering of plies
- Distance between tapes
- Lay-up and scrap strategies
- Sequential order and orientation of courses
- Control of machine and process parameters (pressure, temperature, lay-up speed, compaction, tape tension, etc.)
- Course generation, calculation, and, if necessary, simulation of trajectories (position and orientation of tool center point), collision analysis, determination of the synchronic positioning of all machine axes, and generation of part program
- Other software options specific for each machine configuration and lay-up technology

The processing steps are simulated in order to ensure producibility as well as to avoid defects and collisions between the machine head and the lay-up mold. This step is particularly important with systems employing non-Cartesian, kinematic mechanisms (e.g., robots), where the machine travel/motion is not intuitive and/or not unique.

The set-up of the various lay-up parameters and strategy is strongly dependent on both the lay-up technology chosen and the component geometry. The options offered by the utilized software and the skill of the programmer also play a significant role. Optimization during machine programming has major impact on the following:

- Machine productivity
- Scrap optimization

- Component quality and lay-up performance, in particular for complex geometries and the first ply on the mold (usually with reduced tack)
- Ply staggering over the thickness of the component

The courses and machine parameters graphically defined by the offline software are converted by a post-processor into CNC commands, which can later be interpreted and executed by a specific lay-up machine. The generated machine program (or part program) is imported into the machine control system and executed through a human-machine interface (HMI).

Positioning and Calibration of Mold

During program generation and simulation, the software uses models of the machine, the mold, and the component that represent the assumption of an ideal geometry and position. Once transferred to a real application, there are differences in terms of geometry and position of each element and device in the production line that are created by tolerances set for manufacturing, assembly, and production line. These discrepancies between model and reality may lead to quality problems, non-conformity of the component, and, in the worst case, to collisions between machine and mold.

Therefore, measurement and calibration procedures are used. The positions of predetermined tooling points on the mold are measured and compared to the theoretical, ideal values provided by the CAD model. A transformation matrix can be calculated and applied to the part program so that accurate material deposition and failure-free operation can be assured.

The Lay-up Process and Subsequent Process Steps

The lay-up programs are executed after calibration of the mold. Both lay-up process and parameters depend strongly on the different lay-up technologies. Therefore, the following sections will provide explicit details for both ATL (Section 4.4.2) and AFP lay-up processes (Section 4.4.3).

The result of the lay-up process is an optimized, high-quality component that is accurately positioned in the work area. Therefore, it possible to take advantage of the infrastructure of lay-up equipment to perform post-lay-down operations. For example, cutting tools specifically mounted on the head can be used to trim the component contour.

This approach offers several advantages, such as decreasing component costs by lower capital investment and space requirements, as well as the use of the same reference and position for multiple operations, thus increasing process stability and reducing and simplifying process logistics.

For high-throughput composite manufacturing systems, lay-up and cutting equipment can also be set up in parallel (Figure 4.19). This creates separate zones, in which lay-up and cutting are performed simultaneously within the same system.

Figure 4.19 Parallel cutter and ATL with head change system [Courtesy: M.Torres Diseños Industriales S. A. U.]

4.4.1.3 Prepregs for Automated Lay-up Processes

Unidirectional prepregs consist of oriented fibers that are generally completely impregnated by resin. For automated processes in the aeronautic industry, prepregs with unidirectional carbon fibers and a thermoset resin (typically epoxy) are state-of-the-art.

Independent of specific structural or process related properties, the following main prepreg characteristics are crucial for the automation of the lay-up process:

Tack

Tack describes the property of a prepreg material to adhere to an existing component or mold/form and thus to maintain a desired position and drape after deposition.

Tack influences the lay-up process in a variety of ways, depending on the materials and surfaces encountered:

- Tack between prepreg and mold/form (first-ply issues)
- Tack between prepreg and already laid plies (component)
- Tack between prepreg and paper (ATL) or release film (AFP) prior to lay-up
- Tack between prepreg and guiding elements (e. g., compaction roller or guiding elements in the lay-up head)

As described in Chapter 3, tack is affected by a number of factors (including temperature, compaction and lay-up speed), which can be influenced only to a certain degree during the process. Other factors are strictly material characteristics, which can only be adjusted during the material development phase.

Ensuring accurate process flow may require adjustment of the prepreg tack during the process cycle (e.g., during different stages of material guiding though the lay-up head). In this case, temperature is the easiest parameter to control. Changing prepreg temperature can be used to reduce material tack during material feeding (cooling) or to increase tack at the point of deposition (heating using an activation unit). The rate of deposition and the degree of compaction also influence tack.

Figure 4.20 Material for ATL (left) and AFP (right)

Slitting Quality

Prepregs are typically manufactured in wide rolls (parent rolls) that are subsequently cut to the desired width. The quality, homogeneity, and cleanliness of this slitting process are crucial for optimum lay-up quality, in particular because of the following effects:

- *Tape width and width homogeneity:* Tapes or tows with high variations in width may lead to overlaps or gaps in the laminate as well as to process instabilities.
- *Cleanliness of cut edge:* During the cutting process, filaments, fuzz, or resin build-up may cause contamination of the lay-up head. These in turn may not only cause disruptions in the production process, but they may be deposited on the component and thus cause quality problems.

In general, high-quality components in the aviation industry require tape width tolerances of approx. ± 0.125 mm.

Impregnation

Adequate wetting of the fibers during prepreg manufacturing is important not only for the correct resin distribution during curing. Fiber wetting and resin distribution also impact the lay-up process because they influence tack, cut quality, and the quality of the cutting process both in ATL and AFP processes.

Rigidity and Stability

The stability and inherent rigidity of the material is crucial in order to precisely guide the prepreg through the machine and correctly deposit it. The rigidity of the prepreg ensures that it maintains its geometry during all steps of the lay-up process. Insufficient prepreg stability may result in process instability and in a reduction of lay-up quality.

Prepreg rigidity is particularly important for ATL processes after cutting and right before lay-up once the carrier paper has been removed. Here, insufficient rigidity can cause irregularities during lay-down (Section 4.4.2.3).

For AFP processes sufficient rigidity and stability of the prepreg are crucial because they are processed without stabilizing carrier paper (Section 4.4.3.3).

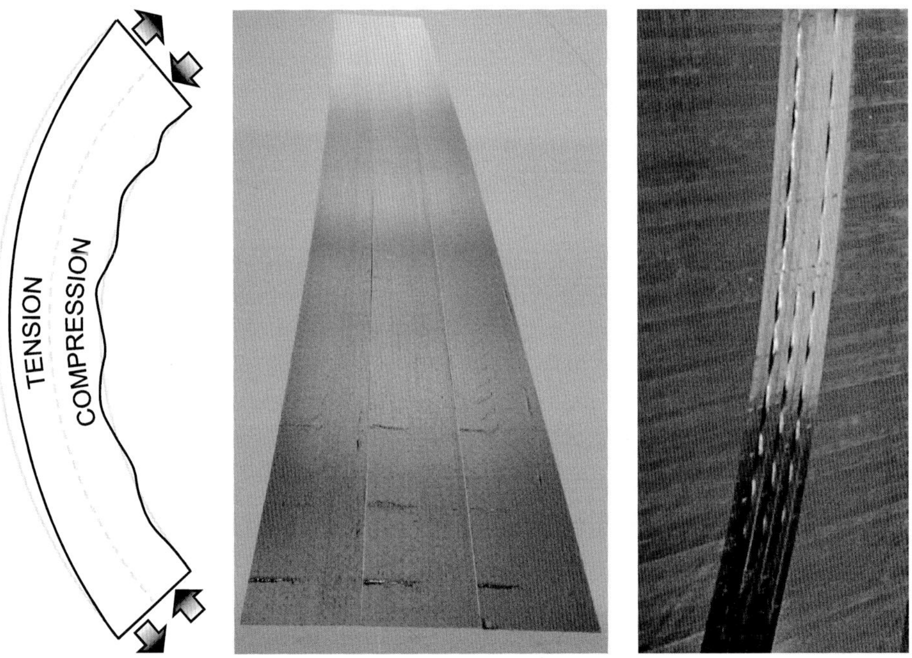

Figure 4.21 Schematic of steering principle (left), ATL lay-up with steering (center), and AFP lay-up with steering (right)

Drapability

The drapability of a tape material describes its ability to follow curvatures in the plane without the formation of buckles.

When laying up (or draping) prepreg tapes along a curve (also called steering), the fibers on the outer and inner radius would have to be of different lengths to allow for a composite part free of residual stresses. Owing to the structure of the unidirectional prepreg and the highly rigid fibers, which do not allow for almost any elongation, the outer side of the tow is under tensile stress, while the inner side of the radius is under compression stress. Therefore, and because of the lower rigidity of the prepreg outside the plane, the residual stresses are released by the formation of buckles/wrinkles on the inner radius (Figure 4.21) and by tow lifting on the outer radius. Both effects are highly influenced by the tack of the tape to the part.

This prepreg characteristic is mostly dependent on the tape width, i.e., it is more difficult to achieve buckle-free lay-up as the tape width increases. However, it is also influenced by other process parameters such as tack, temperature, compaction, and rate of deposition. Narrower prepreg tapes drape easier than wider tapes. Therefore, very narrow tows are used for the lay-up of highly curved components.

Tape Width

The tape width has several opposing effects and is therefore a critical factor for the automation of lay-up processes. The following characteristics are significantly influenced by the tape width:

- *Productivity:* Both lay-up technology and the total lay-up width determine maximum achievable productivity.
- *Drapability of the material:* It decreases drastically with increasing tape width.
- *Flexibility of the machine:* Machines for narrower tape widths are less restricted in terms of material drapability and width of the compaction system and are therefore able to manufacture components with higher complexity.
- *Scrap:* A wider tape causes higher scrap rates, which in turn are also dependent on the lay-up technology.
- *Machine complexity:* It increases with increasing number of sub-tapes or tows.

Prepregs come in standard widths of 1/8 ", 1/4 ", and 1/2 " (AFP), as well as 75, 150, and 300 mm (ATL).

Lay-up head configurations that allow for several simultaneously laid tapes are also available (both for ATL and AFP processes) and can be utilized to optimize the characteristics listed earlier. In ATL-multi-tape configurations, the individual tapes are also called sub-tapes. Here, the total width of the lay-up head is determined by the number of sub-tapes and their width (Figure 4.22).

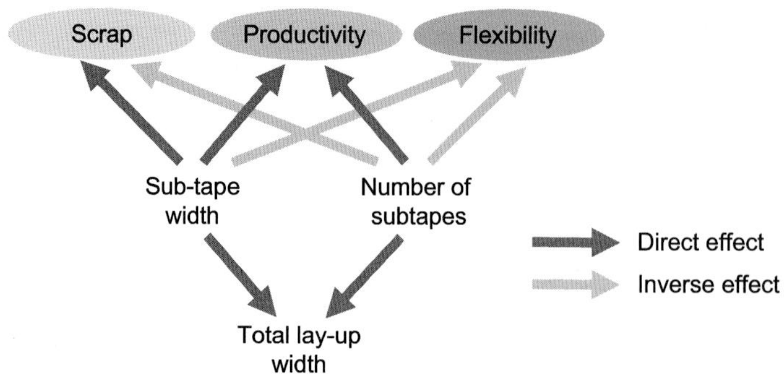

Figure 4.22 Effect of sub-tape width and number on target characteristics

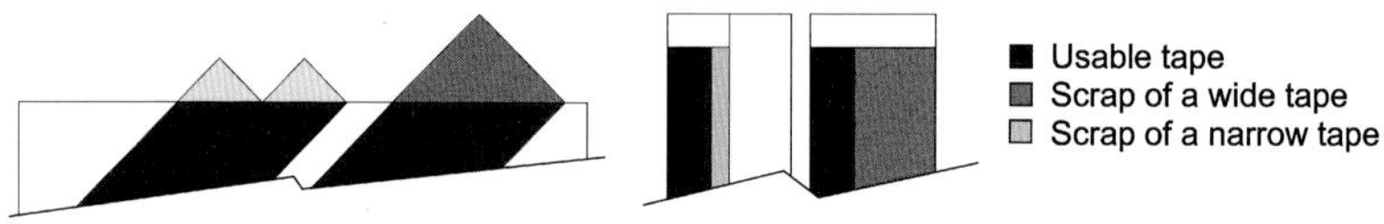

Figure 4.23 Examples of scrap increase with use of wider tapes

Release and Carrier Films

Industrial processes utilize prepreg materials in the form of rolls (for ATL) or spools (for AFP) (Figure 4.20). Because standard thermosetting resins are tacky to a certain degree at room temperature, release paper or film is used to facilitate the unwinding of the prepregs, as otherwise the prepreg tape would adhere to itself. In addition, the release film may play other important roles, depending on the lay-up process used (e.g., in ATL processes, see Section 4.4.2.2).

4.4.1.4 Machine Architecture

A variety of machine configurations can be used to ensure the correct spatial position and orientation of the lay-up head. In order to position the lay-up head perpendicular to the lay-up surface, six CNC axes are typically required (three translational and three rotational axes).

In general, standard machine architectures can be used for all lay-up processes, in particular AFP and ATL. Each machine configuration comes with specific advantages and disadvantages which have to be weighed with respect to the specific application and process requirements. Criteria for the selection of machine configuration include the following:

- Geometry and size of the component (part dimensions and complexity, 2-D or 3-D components, positive/negative mold, ramps, etc.)
- Need for additional positioning axes (horizontally/vertically rotating axis)
- Deposition rate and productivity

- Accuracy and repeatability
- Available space
- Investment, operating costs, etc.

Next, we will describe the machine configurations particularly suitable for ATL and AFP technologies.

4.4.1.4.1 Gantry Configuration

This machine configuration consists of a gantry structure travelling in the machine (longitudinal) direction on parallel rails. A ram bar mounted on the gantry is able to travel in both horizontal and vertical directions. This configuration allows movement along all three translational directions (Figure 4.24, a).

The lay-up head is attached to the ram bar via additional, mechanical, rotating CNC axes. The rotating axes determine the possible orientation of the lay-up head. Depending on application, component complexity, and desired machine flexibility, one to three rotational axes are required in addition to the three translational axes. In general, six axes are required for highly complex 3-D components (e.g., for AFP), while five to six axes suffice for components with medium complexity (e.g., for ATL).

Gantry configurations are suitable for the lay-up of very long and wide components that require highest precision and productivity. Work areas with dimensions reaching approx. 50 × 8 × 2 m can be realized. Gantry type machines in combination with fixed, negative tools are popular in the aviation industry for the lay-up of large components. Examples include wing skins, vertical tails, and fuselage sections, etc. It is also possible to utilize gantry-type machines for the lay-up on positive molds; however, this typically requires the positioning of an additional rotational axis for the lay-up head.

In summary, gantry-type machines provide flexible equipment for the manufacture of large components with very high quality requirements. However, their implementation demands relatively large space and a strong footer (because of the high dynamic loads encountered), which in turn require increased investment costs.

4.4.1.4.2 Bridge Configuration

The bridge-type machine is an alternative to the gantry-type. It consists of a beam resting on statically elevated supports on both sides. Tracks on the supports facilitate longitudinal travel of the beam (Figure 4.24, b).

Here, particular consideration has to be given to material loading operations, because the fixed vertical supports may interfere with lateral loading. Once a material loading concept has been chosen, the bridge configuration may allow for higher machine dynamics in the longitudinal direction than the gantry configuration does, because of its lower moving mass. Safety measures are also simplified because there are no moving machine parts on ground level.

4.4.1.4.3 Column Configuration

The column configuration consists of a column on longitudinal rails. A horizontally attached ram bar can move in both vertical and horizontal directions and is thus able to perform the three fundamental translational motions. The lay-up head is attached to the end of the ram bar via mechanical CNC rotational axes. The rotational axes determine the orientation of the lay-up head; they can be configured to accommodate application requirements (Figure 4.24, c).

Column-type machines are compact and facilitate the manufacture of long, shallow components. In general, they are particularly suitable for lay-up on positive molds (in combination with additional rotational axes). Examples include positive fuselage panels and even complete fuselage sections.

Like gantry-type machines, column configurations provide high accuracy and repeatability and generally require a footer.

4.4.1.4.4 Cantilever Configuration

Cantilever configurations consist of a rail-travelling horizontal beam, on which a ram bar travels in both vertical and horizontal directions (Figure 4.24, d). The lay-up head is attached to the end of the ram bar via mechanical rotational CNC axes. Like the other configurations described here, cantilever-type machines are able to perform each of the three basic translational and rotational motions.

The initial work position of the lay-up head is vertical, and similar to gantry-type machines, cantilever-type machines allow for very long but not very wide or vertical travel of the lay-up head. This machine configuration provides good positioning accuracy and lay-up quality. In general, it requires a machine footer.

This machine configuration can be utilized for long, narrow components on a horizontal table. The cantilever requires less space than the gantry-type machine; however, other than for the gantry, its work area is limited.

4.4.1.4.5 Table Gantry

Other than in the configurations detailed above, in table gantry-type machines the rails for longitudinal travel are not fixed to the floor but directly on the work table, which limits the ability to manufacture wide and highly curvilinear components. However, it has the advantage of a more compact configuration and it requires a smaller footer (Figure 4.24, e).

In general, these machine configurations do not provide the complete six degrees of freedom in motion, but rather only four or five, which allows for the lay-up of flat components with slight curvatures and ramps. It is also frequently used for cutting operations with a six degree of freedom compact cutting head.

4.4.1.4.6 Robot Configurations

In contrast to Cartesian configurations, which are specifically adapted to the requirements of ATL and AFP processes, robot configurations for tape lay-up or fiber placement utilize general industrial robots (Figure 4.24, f). This allows for flexible operations and cost-effective acquisitions. However, because of a robot's limited area of operation, it may be necessary to provide additional axes (e.g., rotational axis, linear axis), which in turn results in additional investment costs.

Standard robot kinematics are generally designed to target points. Other than in Cartesian machines, this may lead to inaccuracies in positioning and orientation during a sequence of set lay-up trajectories. In order to increase lay-up accuracy and thus to meet precision requirements, additional systems or external measuring and alignment processes may be implemented, which also increase the investment cost.

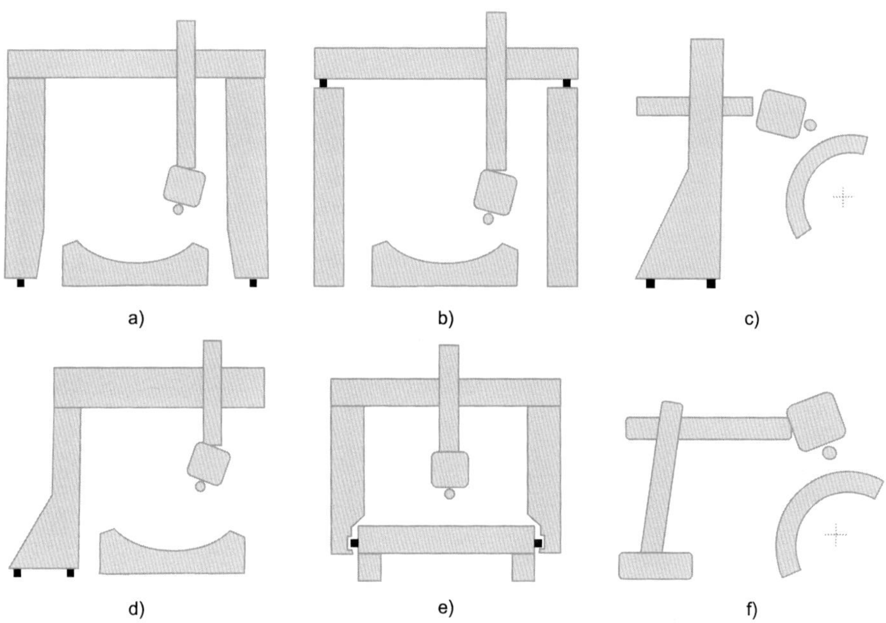

Figure 4.24 Schematic of common machine configurations: a) gantry, b) bridge, c) column, d) cantilever, e) table gantry, f) robot

4.4.2 Automatic Tape Laying (ATL)

4.4.2.1 Introduction

Automated tape laying technology is based on the same principle as manual lay-up methods (Section 4.3) and utilizes sequential, net shape, automated lay-up of wide prepreg tapes. Similar to materials for manual lay-up processes, here prepreg material is deposited on release paper films and wound on cores. The principle steps are also quite similar in both manufacturing processes.

The fundamental characteristics of automated ATL process in comparison to other lay-up processes is the use of one or more wide tapes and the potential to precisely cut and deposit contours. This in turn provides the opportunity to lay up any contour on the component without scrap placed on the part, which constitutes a significant difference to other processes (AFP). However, net shape cutting of wide tapes may cause relatively high tape scrap rates, depending on the component geometry.

ATL processes are utilized for the manufacture of large components with slight curvatures, such as some primary structural components in the aviation industry (wing skins, etc.).

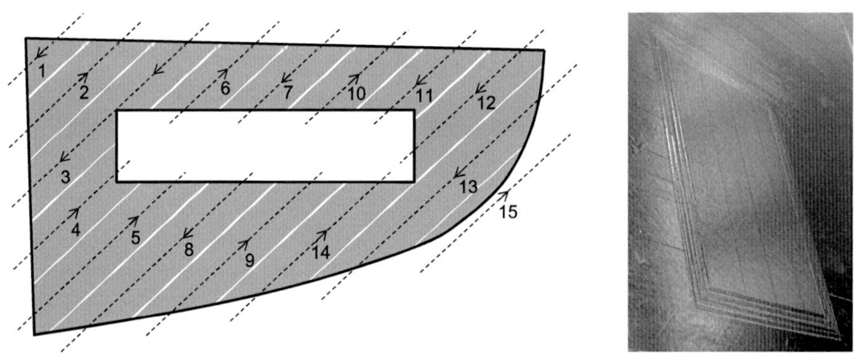

Figure 4.25 Example of course sequence in a net shape ATL lay-up (left) and composite structure consisting of several net shape ATL plies (right)

4.4.2.2 Prepreg Materials for ATL

In general, ATL processes utilize the same material formats as manual lay-up processes (Section 4.3). In addition to the common prepreg characteristics, which influence all automated processes (Section 4.4.1.3), ATL processes require certain specific material properties.

Carrier Paper: Tack and Release

The carrier paper is one of the determining characteristics in ATL processes. In contrast to other processes (e. g., AFP), here the carrier paper allows for net shaped cuts. ATL cutting methods do not damage the paper so that it maintains its supportive function.

Because of the paper's supportive function during processing and in particular during the cutting process, the mechanical properties of the paper, such as stiffness and tear resistance, significantly influence process stability.

The tack between paper and prepreg is another important characteristic of the carrier paper. The adhesion between paper and prepreg is critical because two conflicting properties are pursued simultaneously. On the one hand, tack and adhesion of the carrier paper to the prepreg should be sufficient to support the prepreg (*lamina*), in particular when it is cut. On the other hand, its adhesion should be low enough so that the carrier paper releases easily after lay-up. In addition, the carrier paper has to ensure that the lamina do not adhere to each other on the roll. Therefore, the carrier paper is coated on both sides, although with different types and quantities of release agent. The level of release is a measure for the adhesion or separation/release behavior between paper and prepreg. This interaction between carrier paper and ATL material is shown in Figure 4.26.

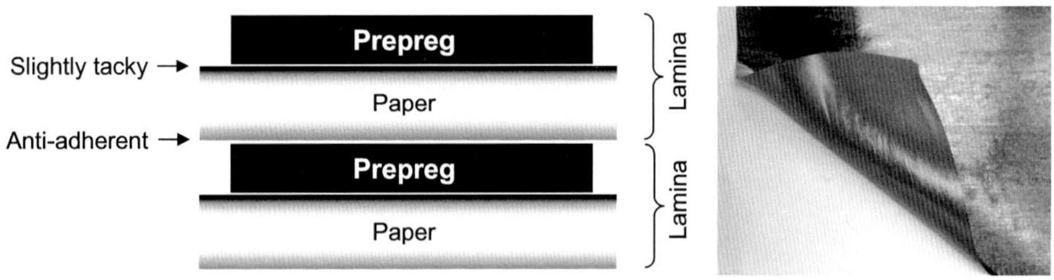

Figure 4.26 Adhesive and separative effect of paper in ATL materials (left), lamina (right)

Tape Width: Drapability, Productivity, and Scrap

The choice of tape width simultaneously affects productivity, scrap, and drapability of the material (Figure 4.22). Wider tapes are suitable for ATL technologies and come in standard widths of 75, 150, and 300 mm. Although it is possible to use narrower tapes, they would reduce productivity.

In general, several tapes can be used simultaneously, which will increase productivity. However, the increasing control complexity and the mechanical design of the lay-up head limit the number of tapes that can be laid up simultaneously.

4.4.2.3 General Principle of ATL Processes

As mentioned earlier, automated ATL processes are based on manual lay-up methods. The process sequences of manual (Figure 4.14) and automated ATL (Figure 4.27) lay-up processes essentially consist of the same steps. However, during automated ATL processes several operations/steps are not only automated, but are typically performed in parallel (measuring, cutting, scrap separation, paper separation, positioning, and compaction). Some steps are also adapted to the automation process in order to facilitate an automated, continuous lay-up process. Figure 4.28 schematically shows the complete process sequence.

Automated processes start with the same prepreg material wound on a roll as manual lay-up processes (Figure 4.20). The precise length and position of the course is measured on-line while the material is unwound. An integrated cutting system is synchronized with the tape and machine movement and is able to make net shaped cuts in the prepreg without damaging the supporting carrier paper.

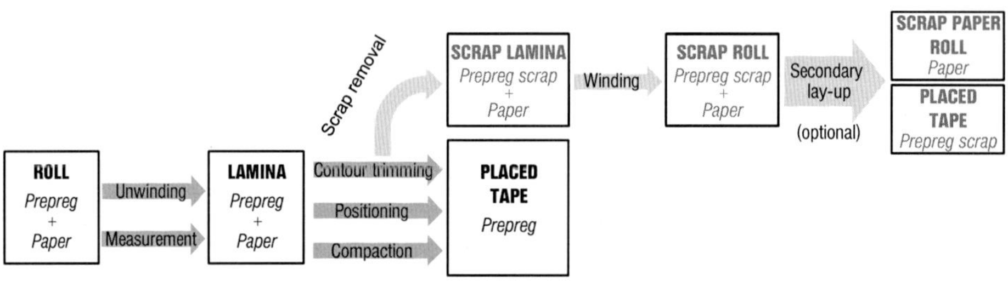

Figure 4.27 Schematic sequence of process steps during ATL lay-up of a prepreg course

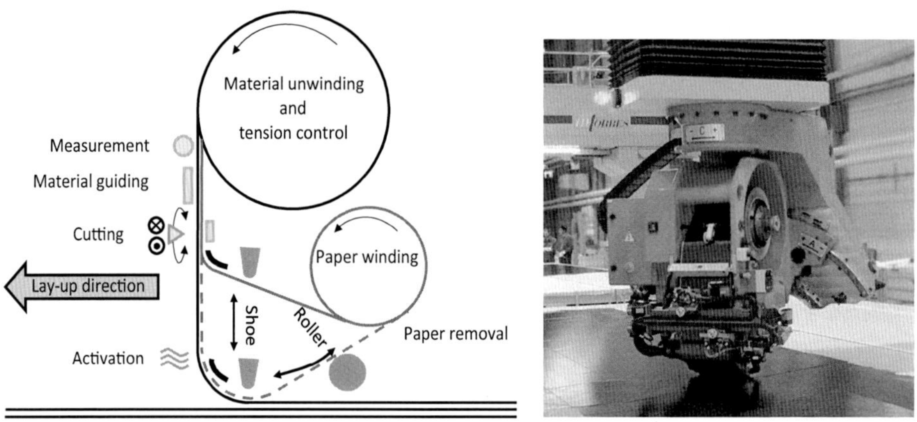

Figure 4.28 Schematic of ATL principle (left) and ATL lay-up head (right) [Photograph courtesy: M.Torres Diseños Industriales S.A.U.]

The lamina now consists of the undamaged carrier paper, the net shape cut prepreg, and the scrap. In the subsequent process steps the net shape cut prepreg is transported to the point of lay-up, then placed on the component or mold and compacted in place. Here it is important to ensure that scrap still attached to the carrier paper remains there and does not contact the component or mold. This is accomplished by dual compaction systems that have to be employed to place complex cuts. The prepreg cut has to exhibit a certain inherent rigidity in order to ensure positional accuracy and lay-up quality.

In order to provide sufficient tack at the beginning of the course, the initial area (approx. 100 to 300 mm, depending on tape width) is actively fed. The carrier paper is then separated from the prepreg by the dual compaction system. It allows for direct contact between the lay-up roll and the prepreg, facilitating the separation of the carrier paper from the prepreg course and the simultaneous take-up of the carrier paper.

As long as the placed prepreg course exhibits sufficient tack and the carrier paper has been separated from the prepreg, the material is no longer actively fed, but rather pulled passively by the travel of the lay-up head. This allows for faster operations during lay-up.

At the end of the course, the net shape final cut is prepared. The cutting system will cut the tape following the set contour in synchronized movement with all other controlled axes.

The generated scrap can be removed together with the carrier paper and later disposed of. If this not desired or if it would interfere with the subsequent process steps, scrap may also be deposited outside the component in the work area during the primary lay-up process. Scrap cuts that cannot be placed outside the component at the end of the courses may be placed in a designated external scrap area in a secondary lay-up process (Figure 4.29). The latter option will result in a moderate reduction in productivity.

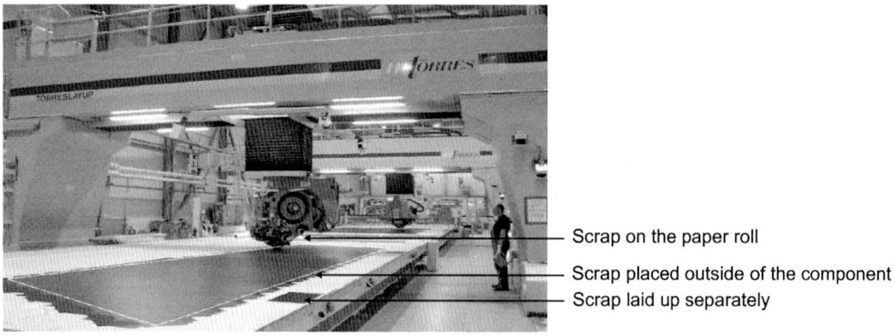

Figure 4.29 Scrap disposal option: on the paper roll, next to the component, or in a designated external area [Courtesy: Airbus Operations GmbH]

In order to achieve highly productive and scrap minimizing processes it is important to tailor the course sequences. With appropriate programming, the final cut of one course can be used as the initial cut for the subsequent course.

4.4.2.4 ATL Technology and Major Sub-systems

Automated ATL processes are rather demanding, both mechanically and in terms of control technology. Prerequisite for a smooth production flow is the optimal interaction between all sub-systems (Figure 4.28).

Material Feed System

The material feed system turns the prepreg roll to feed material, controls tape tension, precisely positions the prepreg, and reels up the carrier paper. The forces and moments that are caused by material inertia and the dynamic loads are managed by robust CNC axes. This also includes measuring systems (typically encoders) that control the exact position and length of the tape at any given point. During different phases of the lay-up process the material feed system is able to move the material both actively and passively.

Guiding System

The guiding system conveys the material from the prepreg roll, through the guiding and the cutting system and down to the lay-up roller. Thereby it reduces friction between material and guiding elements and allows for the compensation of deviations in the reeled tapes (e.g., telescope effects). The system also ensures the correct lateral position of the tapes so that accurate cuts and lay-up can be produced.

Cutting System

One of the most significant characteristics of the ATL principle is fact that it provides the opportunity to lay-up complex contours. This is facilitated by a cutting system that is able to accurately make net shape cuts during lay-up without damaging the carrier paper. Different manufacturers offer a variety of cutting systems.

- *Online cutting:* represents the common ATL process used to realize complex cutting contours during lay-up via multiple CNC axes. Here, the position and angle of the cutters are adapted to the tape's travel. In order to facilitate highly complex cuts, some machines are furnished with two independent cutting units that cooperatively execute the cuts. Here, the prepreg roll is loaded on the lay-up head without prior conditioning. This ensures process stability as it allows for a restart and repetition of a lay-up at any given time. The reduced cutting speed will decrease the productivity slightly. Scrap remains on the carrier paper and can be disposed of in different ways (Section 4.4.2.3).

- *Tape preparation:* Machines with tape preparation (also called 2-stage tape-laying systems) are equipped with rather simple cutting systems that allow for only straight online cuts. In order to lay up complex contours, a second prepreg roll is cut offline following a specific lay-up program. On this second roll, the complex courses are cut sequentially and the created scrap is removed, reducing time required for cutting and scrap take-up during the lay-up process and thus increasing productivity. The relative instability of the lay-up process is a disadvantage of this approach. Because the cuts are prepared in advance, the machine has to be stopped when a cut is laid up inaccurately. In addition, the advanced preparation of the tapes causes an increase in logistical costs and complexity.

Activation

During material guiding, tack should be low to allow for smooth feeding and unproblematic cutting. However, adhesion to the paper has to be sufficient to ensure that the prepreg remains on the carrier during and after the cutting process.

As a general rule, tack is controlled via temperature activation, based on the fact that adhesion increases with increasing temperatures. Standard activation systems include heated air and infrared.

Because tack has to meet opposing requirements during different phases and zones of the lay-up process, the control of tack and temperature relies on compromise. If the activation of the prepreg is too high, the adhesion between prepreg and lay-up roll as well as between prepreg and carrier paper may increase too much, resulting in poor separation and possible process disruptions. If activation is too low it may cause unwanted release of the prepreg from the carrier paper or insufficient adhesion of the prepreg to the component.

Compaction

The compaction system ensures that the prepreg material is correctly positioned and compacted on the component or mold so that sufficient adhesion is achieved while preventing the formation of air entrapment/voids. This is particularly important and challenging when manufacturing complex geometries (e.g., ramps, double-curvature geometries, small radii, etc.) or when utilizing tapes with large overall widths.

Typically, the compaction roller is made of an elastic material that is able to absorb the geometrical curvature of the component/mold. Because the roller is in direct contact with the prepreg, it should exhibit only minimal tack with respect to the prepreg. In order to eliminate contamination of the prepreg, the compaction roller is made of silicone-free materials.

The compaction system also separates the prepreg cut from the lamina and precisely positions it on the component/mold. For complex cuts, a single compaction roller is not sufficient; here, dual systems are typically utilized.

4.4.2.5 Selection of Machine Configuration for ATL Processes

The overall ATL machine configuration is defined by both the lay-up head and the general machine structure. Both factors significantly influence overall performance and have to be adapted to each other and to the component to be manufactured in order to maximize the potential offered by automated processes.

4.4.2.5.1 Machine Architecture for ATL Processes

The machine architecture determines the maximum component size, the component complexity that can be laid up, and the accuracy and deposition rate of the lay-up process. Only appropriate selection and adaptation of the machine configuration to match the specific application allows for the maximization of the benefits offered by lay-up technologies.

Section 4.4.1.4 describes the most important machine configurations in detail. In the following, we will point out some of the distinctive features of ATL configurations.

Gantry/Bridge Configurations

Gantry machines often are the configuration of choice because of their flexibility and high precision. They are popular for large-area, low-curvature components built on negative molds. Among other applications, gantry configurations are used for vertical stabilizers, wing skins, and in general for 2-D components. Depending on component geometry, additional rotating axes can be implemented in order to increase flexibility and to address component complexity.

ATL technology is particularly suited for low to medium component curvature and complexity; here, five CNC axes are generally sufficient. Commonly available gantry configurations are equipped with five or six CNC axes.

Column Configuration

Because of their relatively heavy lay-up heads and material rolls, these configurations are rather the exception among ATL configurations. Nonetheless, they are technically feasible and have been successfully implemented in certain applications.

Cantilever Configuration

Cantilever configurations are best suited for long, narrow components. When used for very wide components, the cantilever design in combination with the relatively high weight of the lay-up head (including the material) may negatively affect the mechanics (static/dynamic) of this configuration.

Table Gantry Configuration

ATL table gantry configurations are suitable for the manufacture of long, narrow, flat components. Here, the longitudinal rails are integrated in the lay-up table, therefore the required footer is small, and the respective investment costs are lower. It is possible to incorporate these configurations in existing buildings with minimal structural alterations.

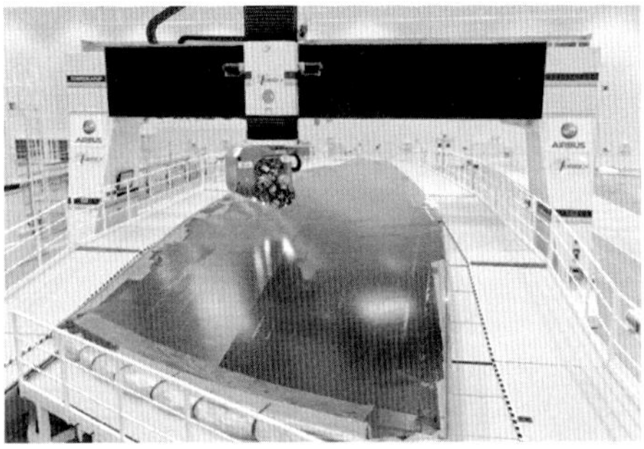

a)

b)

c)

d)

Figure 4.30 ATL configurations: a) gantry-ATL, b) bridge-ATL, c) cantilever-ATL and table gantry cutter, d) robot-ATL [Courtesy: (a, c, d) M.Torres Diseños Industriales S. A. U.; (b) Fives Forest-Liné]

Robot Configurations

These configurations are suitable for the flexible manufacture of components with medium to high complexity. Here, the advantages offered by ATL technologies, such as net shape lay-up, can be exploited.

In general, ATL robots utilize single-tape configurations with narrow tapes (e.g., 75 mm width). These tapes are particularly suitable for applications with higher steering demands. They also lower the weight of the lay-up head, an important factor considering the limited load carrying capacity of the robot.

4.4.2.5.2 Configuration of the ATL Lay-up Head

So far, we described the fundamentals and governing principles of ATL processes and sub-systems. However, ATL technology, like all other manufacturing technologies for composite materials, undergoes continuous development. Newly developed functions and hardware and software upgrades allow for ever increasing productivity and the use of these technologies for more complex components.

Lay-up Head Configuration

Single-tape ATL configurations represent the current industry standard and allow the lay-up of individual tapes. Generally, tape formats range from 150 mm to 300 mm widths. Configurations with high steering demands may also utilize narrower tapes, e.g., 75 mm.

Single-tape configurations offer a high degree of flexibility. However, the selection of the tape width involves a compromise in terms of productivity, component complexity, and scrap. The specific impact can be evaluated only within the framework of an assessment of the intended machine utilization (Figure 4.31, a and b).

Multi-tape ATL configurations are the result of the continuous development of ATL technologies. They facilitate the simultaneous, net shape lay-up of multiple narrow tapes, thus overcoming the disadvantages of single-tape ATL configurations, leading to higher productivity and less scrap. Thus, multi-tape ATL technology represents a combination of the advantages of both ATL and AFP processes.

Commercially available systems generally allow for the lay-up of one (single tape), two, or four tapes (multi-tape).

Both the lay-up head configuration and the relationship between tape width and number of tapes determine the productivity, flexibility, maximum achievable complexity of the component, and the amount of scrap generated by the ATL machine (Figure 4.22). The most important configuration parameters include the following:

- *Tape width:* Wider tapes not only increase productivity and allow for a simple machine configuration, they also increase scrap generation (Figure 4.23). The lower drapability of the tapes decreases the flexibility to lay up complex geometries and ramps. Common tape widths are 75 mm, 150 mm, and 300 mm.
- *Number of tapes (Multi-tape ATL):* ATL application that focus on high productivity, flexibility, and steering as well as on the reduction of scrap utilize a higher number of tapes, in general two or four. A further increase in the number of tapes would increase machine complexity. Common multi-tape configurations include 2 × 75 mm, 4 × 75 mm, 2 × 300 mm, and 4 × 150 mm; they achieve significantly improved results compared to single-tape ATL (1 × 75 mm, 1 × 150 mm, or 1 × 300 mm).

- *Total lay-up width:* Appropriate combinations of tape number and width allow for an increase in total lay-up width and thus for an increase in productivity at constant scrap rates. For example, highly productive multi-tape configurations (2 × 300 mm or 4 × 150 mm) achieve total widths of 600 mm, which far exceeds the standard width of 300 mm for single-tape configurations. However, for components with certain curvatures the corresponding width of the compaction roller may lead to insufficient compaction (e.g., of ramps).

Multi-tape configurations with total widths of 300 mm (2 × 150 mm, 4 × 75 mm) are typically used for components with moderate curvatures and high productivity requirements at low scrap rates, e.g., wing skins (Figure 4.31, c).

Highly productive multi-tape configurations with total widths of 600 mm (4 × 150 mm, 2 × 300 mm) are used for large area components with relatively marginal curvatures and high productivity requirements, e.g., horizontal stabilizers (Figure 4.31, d).

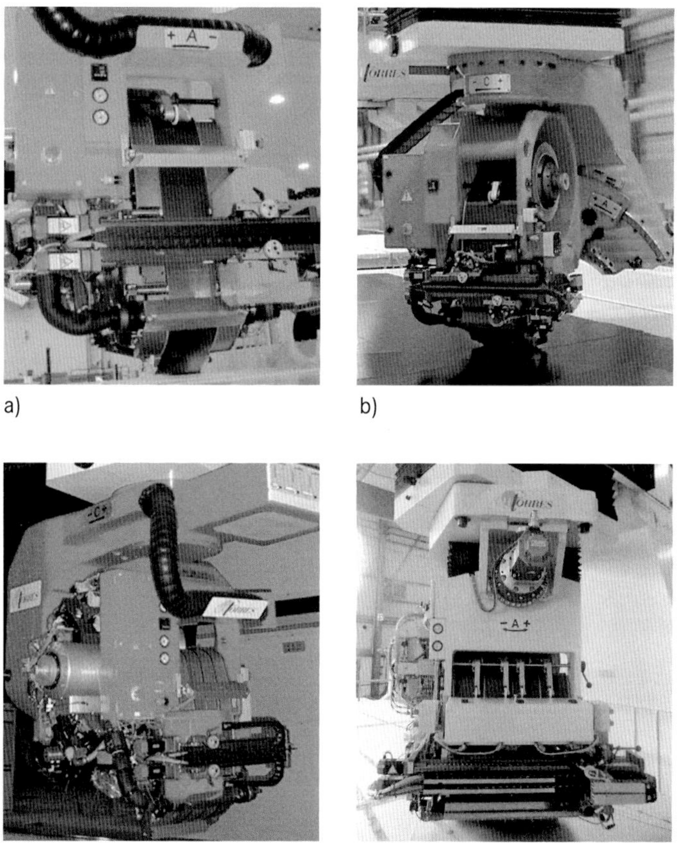

a) b) c) d)

Figure 4.31 Lay-up head configurations: a) Single tape 1 × 150 mm, b) single tape 1 × 300 mm, c) multi-tape 4 × 75 mm, d) multi-tape 4 × 150 mm [Courtesy: M.Torres Diseños Industriales S.A.U.]

Optional quick change lay-up head configurations are available. They facilitate not only quick material substitution, but also offer the option to optimize lay-up of specified component areas with varying requirements by using tapes with different widths (Figure 4.19).

The lay-up head configuration should be carefully selected and optimized for each specific application.

Lay-up Technology

Commercially available technologies may offer substantial differences in advantages and disadvantages with respect to a specific application. Therefore, careful analysis is required when identifying the best suited technology for a given application. The following criteria should be considered when selecting an ATL technology:

- **Planned production:** Precise identification and definition of the components to be built, their geometries, requirements to be met in terms of productivity and quality, prepreg materials to be used, as well as of all other pertinent conditions are essential to making a suitable, sound selection of the optimal ATL technology for a given application.
- **Machine structure:** Determination of suitable machine configuration and optimum number of main axes (linear and rotational). Machine configurations with four main axes are used for flat lay-ups without the option to lay up pronounced ramps. Five CNC axes allow for complete 3-D functionality. The full array of six CNC axes makes it possible for the lay-up head to be continuously positioned perpendicular to the (component/mold) surface. This option is only necessary for extremely complex components, for which ATL technology is typically not utilized.
- **Lay-up head configuration:** Choice between single- and multi-tape configurations, and option of quick change lay-up head system.
- **Cutting system:** Using flexible, reliable, but time-consuming online cutting systems, or more productive but high-cost and less stable tape preparation configurations.
- **Other options:** Other options and systems can often be implemented in the lay-up head. They provide added features for quality assurance, process stability enhancement, or productivity increase, both during the lay-up process and over the course of the entire production process.
  - *Contour cutting systems:* Additional cutting systems facilitate the trimming of the laid-up component to its final contour. They significantly reduce capital investment, space demands, and logistics, and lead to quality improvement because component lay-up and trimming take place under identical machine references and component positioning conditions (Figure 4.19).

- *Component labeling, contour marking, and laser control of contour:* facilitates the labeling and/or laser projection of the ply contour on the mold. Often used for the identification of cuts, for the positioning of manual plies, or for the control of lay-up and cut contours.
- **Costs:** capital investment (costs of machines, infrastructure, and footers) as well as operative costs and additional systems should be considered in terms of productivity and machine utilization.

Depending on application, additional criteria may have to be considered.

4.4.3 Automatic Fiber Placement (AFP)

4.4.3.1 Introduction

Automatic fiber placement technology is based on the lay-up of narrow tows that are laid simultaneously and independently of each other. They can be cut by a straight, transversal cut. Thus, the prepreg is not net shape placed, but rather near net shape (Figure 4.32). Each tow and its position and length can be controlled individually, so that there is almost no excess length at the component edges, which reduces scrap to almost zero. In order to achieve high productivity, up to 32 parallel tows are placed simultaneously. Here, the principles of operation are entirely different compared to manual lay-up or automated tape laying methods (Section 4.4.2), which precisely lay up wider tapes following a given contour.

While the drapability of prepreg tapes depends on their respective width, the narrow tows used for AFP processes increase the drapability of the material and thus facilitate the manufacture of highly complex 3-D components. In addition, AFP processes reduce scrap because the individual tows are precisely controlled and cut.

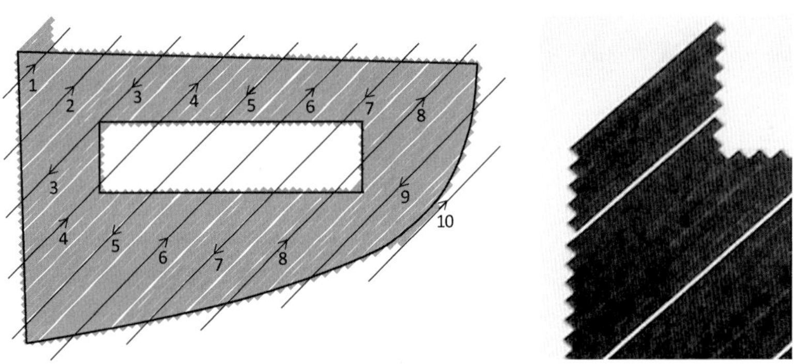

Figure 4.32 Example for ply contour and course sequence with AFP (left), AFP lay-up with typical cut patterns and minimal lay-up length (right) [Courtesy: right - Fraunhofer Institut für chemische Technologie, ICT-FIL]

In the past, AFP technology was characterized by reduced speeds during cutting or during adjustment of the configuration. Today, new developments, such as cutting and adding at full lay-up rates (so called cut/add on the fly), have opened up a whole range of new opportunities and led to significant increases in productivity.

4.4.3.2 Prepreg Materials for AFP

AFP processes utilize prepreg slit-tapes, which are narrow fiber tows in standard widths of 3.175 mm (1/8"), 6.35 mm (1/4"), and 12.7 mm (1/2") that are manufactured by slitting a parent tape. Similar to ATL tapes, slit-tapes come in single-ply thicknesses of 0.125 mm (134 g/m^2), 0.184 mm (194 g/m^2), and 0.25 mm (268 g/m^2), depending on application. These fiber tows are also trimmed to specific widths (slitting) by rotary knives from so-called parent tapes and subsequently wound on smaller spools of 3 kg, 7 kg, or 11 kg.

In addition to the general characteristics of thermoset prepregs for all automated processes mentioned in Section 4.4.1.3, AFP processes require important material properties specific to these processes.

Plastic Release Film

In general, processing of slit-tapes does not require a carrier material. However, thermoset prepregs for AFP processes are provided with a plastic release film, e.g., a PE film (Figure 4.33). It prevents the thermoset material, which is tacky at room temperature, from sticking to itself. In contrast to ATL processes, here the release film does not act as a carrier material; it is directly removed in the creel house and the material is fed through the machine without additional films. The tow feeding mechanism is often cooled in order to reduce the material's tack through the machine elements.

Figure 4.33 Slit-tape (6.35 mm) on spool

Tack

The tack of the prepreg material is of even higher importance for AFP technologies than for ATL processes, because AFP component geometries are often complex (where steering creates higher inner tension) or the tows have to be laid up on components that rotate around an axis, on which the tapes are subject to gravity. Particularly critical is the first ply tack.

Material tack can be controlled by the targeted selection of activation temperature, compaction, and lay-up rate. In extreme cases and for specific applications, the use of adhesive films may be necessary.

Tow Width: Productivity, Drapability, and Scrap

The drapability of a prepreg material is not only dependent on its composition (fiber and resin) and the process parameters (compaction, temperature, lay-up rate), but primarily on its width. Therefore, the use of wider tow formats leads to a reduction in drapability and thus in process flexibility because lower drapability reduces the range of component complexity. It also creates higher scrap rates (Figure 4.23).

Fiber placement heads for AFP processes generally use tows with standard widths of 1/8" (3.175 mm), 1/4" (6.35 mm), and 1/2" (12.7 mm), and a standardized number of tows (8, 16, 24, or 32).

Slitting Quality

During AFP processes large numbers of tows are combined, and therefore compliance with tape width tolerances is essential in order to avoid gaps and overlaps.

High quality components in the aviation industry typically allow for tape width tolerances of ± 0.125 mm. A good slitting quality leads to enhanced cleanliness on the tape guiding.

4.4.3.3 Governing Principle of AFP Processes

Automated AFP processes simultaneously deposit several narrow, pre-impregnated tows next to each other. The individual tows are controlled independently so that they can be started and cut as required (Figure 4.32). The complete process is schematically shown in Figure 4.34.

The prepreg material is provided on spools, which are placed on the creel. The spools are unwound as the tows are paid out and the release film is removed. Subsequently, the tows are guided through the machine up to the placement head under controlled tow tension.

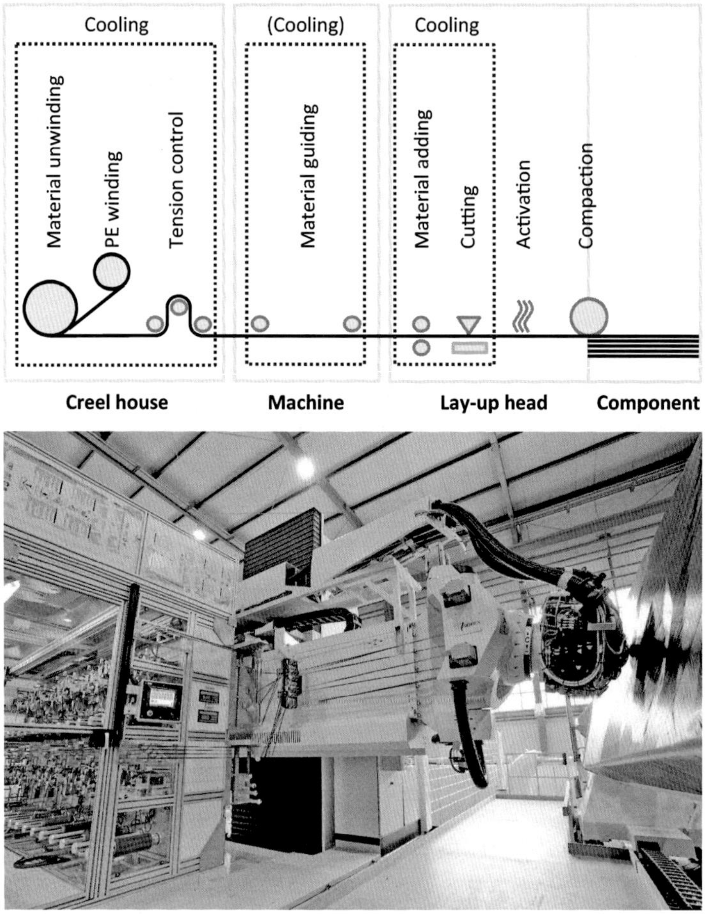

Figure 4.34 Schematic depiction of AFP sub-processes and their position within the AFP line (top) and typical AFP processing line (bottom) [Courtesy: M.Torres Diseños Industriales S. A. U.]

Because the thermoset material is tacky at room temperature, the tow guiding elements may need to be cooled. The cooling system reduces prepreg tack and thus allows for better feeding of the material through the machine and it reduces contamination and resin/fiber accumulation.

Within the fiber placement head the individual tows may be actively fed if required. In addition, the tows are cut transversally by an integrated and CNC-controlled cutting system. Immediately prior to lay-up, the tows are deposited in parallel. In order to ensure adequate tack of the prepreg on the component, an activation system increases the temperature of prepreg (and/or component surface). Subsequently, the tows are laid up with the help of a compaction roller and draped on the component surface or mold (Figure 4.35). A certain inherent stiffness of the material is necessary in order to guide the tows in the correct position under the roller.

All automation steps, the positioning, and the synchronized movement of all CNC axes are part of the machine program and are controlled by a CNC/PLC. Individual AFP machine manufacturers provide proprietary solutions and sub-processes (some of which are patented), which may result in slightly different processes compared to the general process described here. Additional details regarding the mentioned sub-systems can be found in Section 4.4.3.4.

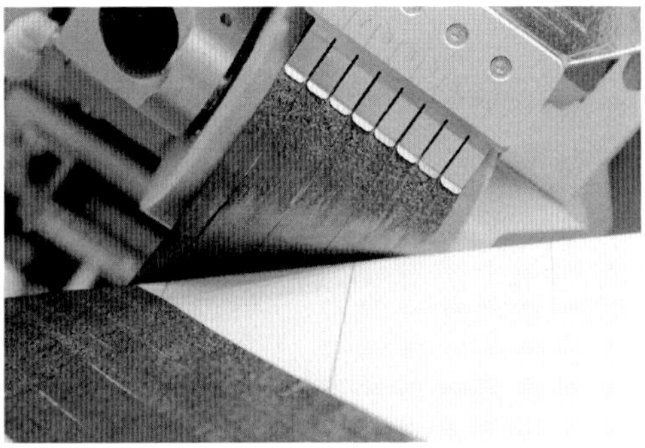

Figure 4.35
Parallel deposition of tows, activation, and compaction [Courtesy: Fraunhofer Institut für chemische Technologie, ICT-FIL]

4.4.3.4 AFP Technology and Relevant Sub-systems

Selection and optimization of the machine configuration are crucial in order to achieve the desired level of component quality and productivity. The most influential factors to consider during the selection of a machine configuration are the choice of placement head technology and configuration as well as the overall machine architecture. In an optimal scenario, they would be selected for the specific component to be built.

Creel: Material Unwinding, Release Film Removal, and Tension Control

The material spools are stored in the creel, which is generally cooled to maintain low tack and enhance material tack life. The spools are individually driven to maintain exact tow length and tension required for the lay-up process at any given time. In order to achieve this, the spools are equipped with a tow tension controller that allows for a controlled state of tow tension in the machine and thus for an accurate process flow, which in turn positively influences lay-up quality and process stability. In addition, the creel contains a system that removes the release film from the tows. This system may wind up, vacuum, or shred the separator film.

The creel may be placed in different positions; however, size and weight of the system are limiting factors. It is necessary to find a compromise between guiding complexity until the tows reach the placement head and the geometrical interference and inertia of the creel. If the creel is placed at a larger distance from the placement head, material guiding becomes more complex. On the other hand, its travel and inertia are reduced. If the creel is placed closed to the placement head, the guiding becomes easier; however, it leads to inertia, dynamic loads, and larger structures.

The creel can also be directly connected to the placement head. Here, the distance between creel and placement head is small, constant, and independent of machine travel. This significantly simplifies the design of the guidance system and facilitates the design of *head change configurations*, thus providing for quick changes of material and head configurations (tow number and width). This configuration is most interesting for high-volume production and highly variable laminate complexity applications. However, the heavy and rather voluminous creel structure, which is now directly attached to the placement head, may cause geometrical limitations, which in turn may affect productivity or interfere with the ability to produce certain components.

There is also a possibility to place two material sets on the creel. An *autosplicer* system splices and automatically switches from the empty set to the new one once the first is depleted. This option is of particular interest for high-volume productions in order to minimize material loading time.

Material Guiding System

The material guiding system drives the tows, without separator film, from the creel, through the placement head, to the lay-up roller. In general, the material guiding system consists of guide pulleys or tubular guides, which compensate for the relative movement and rotations of the placement head with respect to the creel (Figure 4.36). Here, the goal is to prevent rewinding of the spools caused by the yanking motion of the machine, which would create process instabilities.

a) b) c)

Figure 4.36 Creel and feeding systems: a) by guide pulleys, b) by tubular guides, c) creel and feeding system at the placement head [Courtesy: (a, c) M. Torres Diseños Industriales S. A. U.; (b) Coriolis Composites S. A. S.]

In addition, the tows should be fed steadily in order to prevent tow twist. This is of particular importance for materials with relatively low intrinsic stiffness. A certain level of intrinsic stiffness is necessary, not only to ensure accurate feeding through the machine, but also in order to correctly position the tows from the placement head on to the roller, because after exiting the placement head, the tows are conveyed to the compaction roller without any mechanical guiding elements.

Cooling

If required, the material can be actively cooled through the guiding system. The cooling system reduces the tack of the prepreg and thus facilitates better material feeding, an enhanced cleanliness of the tape guiding, and and higher process stability. This is mandatory for systems with closed tubular guiding, in which prepregs and guiding system share large contact surfaces.

Feeding System

The feeding of the material during the lay-up process changes from active to passive. At the beginning of the course, the material has to be actively fed until it is in contact with the component and a minimum length has been deposited so that sufficient adhesion between tow and component has been achieved. Subsequently, the tack of the already deposited material is able to overcome the tape tension and material is simply pulled by the machine travel. Thus, the material is passively conveyed. In general, active feeding systems are based on rotating elements that allow for the independent, controlled conveyance of individual tows.

Position control and accuracy of the feeding system affect lay-up accuracy when adding tows. This is particularly critical at high deposition rates when inaccurate feeding technology and control may create unacceptable tolerances.

Cutting System

The cutting system is crucial for overall productivity and lay-up accuracy. Here, the differences between individual AFP technologies are significant, caused by the different (usually patented) approaches by the machine manufacturers.

State-of-the-art cutting systems for industrial AFP processes using thermosetting materials use straight cuts – transversal to fiber orientation – for each individual tow. The result is the typical pattern of AFP laminates (see Figure 4.32). Highly complex cuts, such as the ones that can be achieved using ATL technologies, cannot be performed, because the material is not supported by carrier paper during cutting. Thus, only *near net shape*, not net shape lay-up is possible. The discrepancy between theoretical and actual lay-up depends on the tow width.

The designs and control routines set by the machine manufacturers significantly influence the main characteristics of the cutting system, including the minimal lay-up length, maximum cutting speed, and the cut-restart length once the tow has been cut.

The *minimal lay-up length* is determined by the mechanical distance between cutting system and lay-up point. Depending on technology, this length may vary between approx. 90 and 140 mm and it influences the adaptation of the subsequent lay-up to the theoretical contour, in particular for complex contours. Very narrow contours and/or contours with corners can be laid up only by laying scrap (Figure 4.32). If possible, scrap should be positioned outside of the component. However, sometimes laying scrap inside the component is unavoidable.

The *maximum cutting speed* is a critical criterion, determining the productivity of AFP technology. Simple cutting principles, such as the *Guillotine principle* [3], require reduced cutting speeds (approx. 0.2 to 0.4 m/s). Highly developed principles, such as the *rotational principle* [1], allow for cutting operations at full speed (cut/add on the fly), which is typically 1 m/s. For rotational cutting systems, the cutting accuracy is typically higher, in particular at high lay-up rates.

The cut-restart length is the minimal length necessary to lay up a tow after the previous was laid down. It is dependent on the configuration of the cutting and feeding systems.

Today, it is also possible to cut individual tows at variable angles for specific applications in order to achieve a more accurate lay-up [2]. However, for the current AFP production machines for thermosetting materials this is not yet state-of-the-art.

Activation

The tack of prepreg materials is strongly dependent on temperature. In order to increase adhesion of the prepreg on the component, the tows' resin (or the component surface) is lightly activated by heating systems. Activation is particularly necessary in systems with cooled material feeding, which cool the material to approx. 10 to 12 °C.

Depending on system configuration and geometry, either the material to be laid up or the already laid-up material (component or mold) or both can be activated.

Standard activation systems for thermoset prepregs are hot air and infrared lamps. Hot air systems are often difficult to control. Infrared lamps on the other hand offer fast and accurate control.

Compaction

The compaction system generally consists of an elastic roller that can adapt to the shape of the component by a compression mechanism. It drapes the tows over the component and ensures their accurate position and sufficient adhesion between tows and component. It must also avoid air entrapment while draping the material.

Component geometry (double curvature surfaces, ramps, etc.), compaction force, as well as geometry, design, and material of the roller substantially influence compaction quality. In order to provide optimum compaction across the entire tape width, the roller width (and thus the number and width of the tows) have to be adapted to the complexity of the component. Very wide and rigid rollers will not be able to sufficiently compact highly complex geometries with ramps or strong curvatures, which may lead to the so-called *bridging effect*.

4.4.3.4.1 Machine Architecture for AFP Processes

Machine architecture determines the maximum component size, the achievable component complexity, and the accuracy and rate of deposition. The correct choice of machine architecture therefore has major impact on the future productivity of the AFP system. Section 4.4.1.4 provides a detailed description of the major common machine configurations. The following will detail some specific characteristics of these configurations for AFP applications.

Gantry/Bridge Configuration

Gantry and bridge configurations (Figure 4.37, a) are often used for AFP machines because of their flexibility and highest level of precision, in particular for large-area components and flat or negative molds without a rotational axis (e.g., negative fuselage skin molds). All six degrees of freedom of the placement head are used for components with medium to high curvatures. In combination with a rotating axis (mandrel) it is possible to lay up complex positive molds.

In this configuration, the position of the creel has to be considered as it affects the length and complexity of the material feeding system, the machine dynamics, and creel accessibility.

Column Configuration

The column configuration (Figure 4.37, b) exhibits a rather compact architecture with a relatively simply material feeding system as the creel is close to the lay-up head. It provides high levels of flexibility for a wide variety of components. Therefore, it is often utilized with AFP technologies, in particular in combination with an external, rotating axis. This configuration is particularly suitable for the manufacture of panels, positive fuselage panels, entire fuselage sections, or spars.

Cantilever and Table Gantry Configurations

Cantilever (Figure 4.37, c) and table gantry configurations are particularly suitable for narrow, long components. Novel developments for cantilever AFP machines, in particular cut/add on the fly, increasingly allow for the manufacture of planar components with a large number of recesses. Here, the ratio of scrap is significantly reduced in comparison to traditional ATL technologies. The simultaneous increase in productivity and reduction of scrap is possible only because of the mentioned "cut/add on the fly" technology, which can be used for the manufacture of planar laminates, such as stringers.

Robot

Despite their limited accuracy and dynamic with respect to complex tool center point movements and work areas, robots can be utilized for the manufacture of complex components (Figure 4.37, d). They are of particular interest for research and development applications because of their flexibility, simple design, and relatively low costs.

Highly complex components can only be laid up in combination with additional axes, such as a linear translation of the robot or a mandrel, which in turn increases machine complexity.

Narrow placement head configurations for 8 or 16 tows are standard for robotic AFP applications. These small configurations are best suited for the targeted components and the reduced payload of the robot.

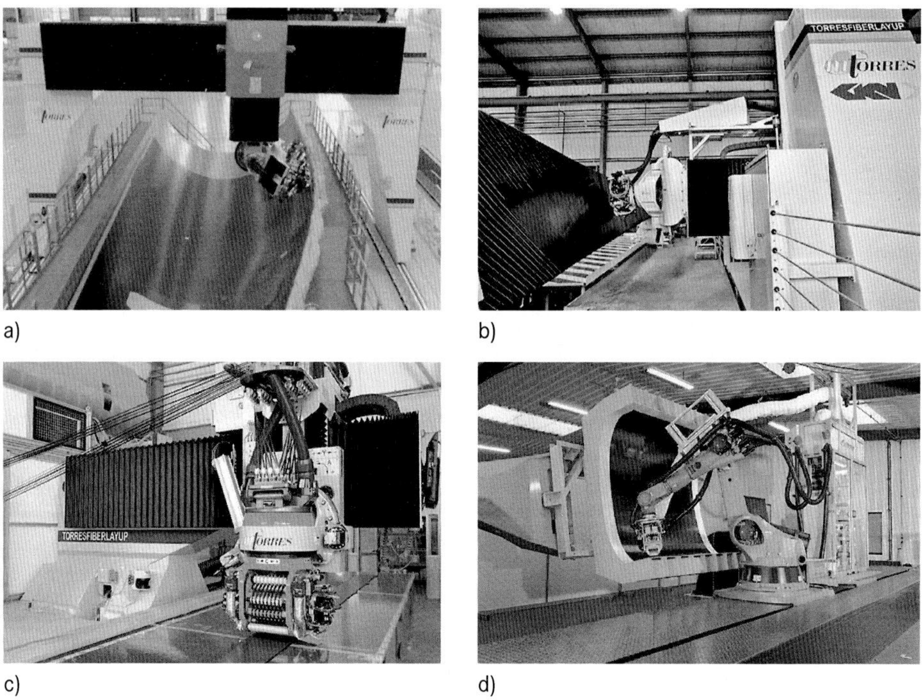

Figure 4.37 Common AFP configurations: a) gantry, b) column, c) cantilever, d) robot [Courtesy: (a) Premium Aerotec GmbH; (a, b, c) M. Torres Diseños Industriales S. A. U.; (d) Coriolis Composites S. A. S.]

4.4.3.4.2 Design of AFP Placement Heads

The configuration of the fiber placement head has to take into account the general characteristics (such as tow number and width), but also the placement head technology. The geometry definition of the components to be built is critical for the optimization of this configuration.

Fiber Placement Head Configuration

Fiber placement head configuration and the interaction between tow width and number determine the flexibility and productivity of the AFP placement head (Figure 4.22). The most important configuration parameters include the following:

- *Tow width:* Wide tows increase productivity and match the simple design of the machine; however, they also increase scrap and contour inaccuracy of the lay-up. Their low drapability limits the lay-up of complex geometries (such as components with strong curvature and ramps). Common tow sizes are 1/2", 1/4", and 1/8".
- *Overall lay-up width:* Productivity can be increased by increasing the overall lay-up width (by either increasing tow numbers or width). However, the correspondingly wider compaction roller will create more geometrical interferences between placement head and mold, in particular in more complex components. This in turn will result in problems with compaction and will limit the range of component geometries.
- *Number of tows:* Applications focused on high flexibility and steering capability utilize a reduced number of narrow tows (8 to 16 tows of 1/8" or 1/4" width). Components with moderate levels of complexity and high productivity requirements are better served with a higher number of wider tows (16 to 32 tows of 1/4" or 1/2" width).

Therefore, the placement head configuration (tow width and number) should be carefully optimized for each specific application in order to achieve the desired goals in terms of productivity, component quality, and complexity.

Lay-up Technology

Although the described principles of AFP technology are universally valid for all AFP applications, there are different fiber placement systems available on the market. Some of those come with exclusive, patented systems, which may differ significantly in terms of the used lay-up technology (Section 4.4.3.4). They considerably affect the flexibility of the machine, the quality and accuracy of lay-up, and the productivity of the AFP process.

The following criteria should be considered during the selection of an AFP technology:

- **Intended production:** Crucial for the informed selection of the optimum AFP technology is the clear definition of the component to be built, the component geometries, productivity and quality requirements, prepreg materials to be utilized, as well as other, general restrictions.
- **Creel:**
  - *Creel position:* Affects accessibility, overall material feeding, and, because of its voluminous structure, the overall machine design.
  - *Standard creel, autosplicer configurations, or placement head change systems:* Depend on planned production and time needed for material loading.
- **Subsystems:**
  - *Cutting and feeding systems:* Highly productive "cut/add on the fly" systems operating at full deposition rates (approx. 1 m/s) and high accuracy during adding and cutting, or more basic, space-saving, but significantly less productive guillotine-type configurations.
  - *Bidirectional lay-up:* Describes the option to lay up a component alternating in both directions by either quickly rotating the placement head or by switching the position of the lay-up roller. This approach avoids long positioning paths; it is of particular interest for the lay-up of large components.
  - *Activation system:* It should be able to attain sufficient tack and activation in the material to be laid up without limiting the rate of deposition. In addition, it needs to provide appropriate control under varying process parameters.
- **Costs, supplemental systems, and expansion options:**
  - *Capital cost:* Capital expenditures, infrastructure and footer costs, among others, should be balanced with respect to productivity and utilization.
  - *Operative cost:* Consumables, energy, cooling, and air use, etc., as well as maintenance costs, must be evaluated and balanced depending on the type of AFP technology.
  - *Personnel cost:* Here, the costs required to operate the machines need to be considered. They include not only the operator costs, but also the costs for programming, which depend on the selected AFP technology and software, as well as other indirect personnel costs.
  - *Cost for supplemental systems:* These include the systems required to meet production requirements. For example, systems to increase positioning accuracy (such as laser measuring systems) or additional controlled axes (rotating axis, etc.) for lay-up operations.
  - *Expansion options:* Effects of expansion systems on overall performance (accuracy, productivity, etc.) and their respective costs.

- **Operator convenience:**
  - *Software for offline programming and simulation:* The software is an important factor to be considered during selection of an AFP system. It determines programming time, time required for optimization and test of new components, component quality, and productivity. Technically mature software and an optimized part program will fully exploit machine capacities and facilitate optimal results in terms of resource utilization.
  - *Maintenance and material loading:* Simple material loading and maintenance/cleaning systems reduce machine down times.

Depending on the specific application, additional criteria may also have to be considered.

In general, machine configurations can be individually tailored to meet customer requests.

4.4.4 Productivity and Cost Effectiveness of Lay-up Processes

In order to determine the best suited lay-up process, an adequate economic evaluation is crucial. Among others, the lay-up productivity is a decisive parameter in the evaluation of prospective lay-up processes. It requires the comprehensive analysis of all subsequent manufacturing steps along the critical path of machine utilization (e.g., assembly, calibration and disassembly of the mold, component transfer) as well as the recurring costs and required capital investment for the entire lay-up process. The respective cost structures, desired production rates, as well as the considered lay-up technology and machine design will significantly influence this analysis.

4.4.4.1 Lay-up Productivity

The comparison of equipment and technologies requires the determination of the lay-up performance of these processes. End users in particular often express this productivity in terms of kg/h. However, it is not realistic to make a general statement regarding productivity in these terms, because it depends on a variety of factors, including the following:

- Lay-up width, machine configuration, and lay-up technology
- Material and areal weight
- Component specific data (component complexity and size)
- Programming and lay-up strategies

In order to allow for a better comparability and analysis of the net machine and technology capacities, the average lay-up velocity (ALV) was introduced as a suitable unit. The following section also addresses its relationship to other common productivity units.

Average Lay-up Velocity (ALV):

In general, the common lay-up technologies come with a maximum lay-up velocity of approx. 1 m/s. Therefore, this value will be used as a benchmark. However, specific machine configurations allow for higher maximum velocities, although this maximum performance will be achieved for only certain phases of the component lay-up process. The average lay-up velocity of the entire process is significantly lower because of a wide variety of delays and productivity reductions, e.g., caused by short courses and/or curvatures, empty runs, transition movements, and others.

In the following we will introduce the average net lay-up velocity, measured in net linear meter per second (nlm/s), which allows for the determination of absolute machine capacities. Net ALV considers not only the lay-up velocity, but also additional process operations during lay-up that significantly reduce productivity:

1. *Reduced process velocity:*
   - Acceleration and deceleration phases during lay-up, e.g., tape beginning and end
   - Cutting systems: reduced cutting velocity when performing net shape cuts (ATL), or with guillotine-style systems (AFP), etc.
   - Process- and material-related velocity reduction; for example because of high geometric complexity, limited tack, extreme drape, lay-up quality, etc.
2. *Non-productive time:*
   - Process-related wait times (SPS/CNC cycles, non-time-optimized processes)
   - Non-productive machine travel, e.g., because of time between tapes, sequences, and additional time caused by non-optimized programming
3. *Non-productive process steps:* e.g., lay-up of scrap on external scrap area (ATL)

In addition, the ALV is largely independent of lay-up width, material, and areal weight. Here, only the component complexity and size, the lay-up technology and parameters, as well as the process itself exert relevant influence. This even allows the comparison of the net lay-up performance of different processes (ATL vs. AFP) and lay-up technologies (machine manufacturers) for the same component.

Figure 4.38 is a qualitative depiction of ALV as a function of average lay-up length for different lay-up technologies and component complexities.

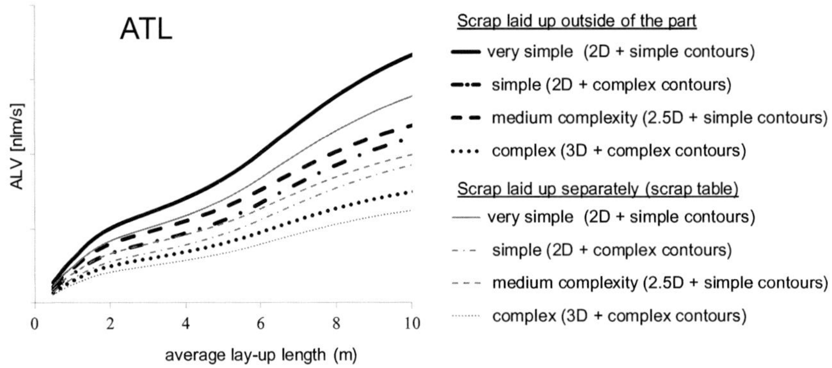

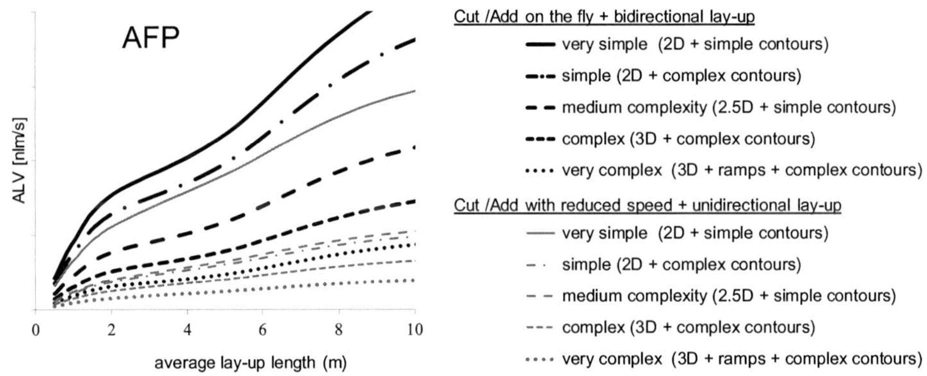

Figure 4.38 Qualitative values of ALV (nlm/s) for various AFP and ATL technologies and component complexities

Areal Lay-up Rate (m^2/h):

The areal lay-up rate (typically measured in m^2/h) is dependent on the ALV (nlm/s) and the overall lay-up width. The numbers and widths of tows/tapes, thus determine the lay-up width per course (Figure 4.39).

Therefore, the areal lay-up rate is considered a reasonable measure for the comparison of different placement head configurations. Please note that lay-up width and material may affect lay-up velocity.

Productivity in Terms of Mass (kg/h):

For each specific application and prepreg material, the areal lay-up rate and the specific material weight can be used to calculate the productivity in terms of material mass kg/h.

	Tape / Tow width	Total lay-up width of typical machine configurations (mm)						
		Number of tapes or tows						
		1	2	4	8	16	24	32
ATL	75 mm	75 mm	150 mm	300 mm				
	150 mm	150 mm	300 mm	600 mm				
	300 mm	300 mm	600 mm					
AFP	1/8"	3.175 mm	6.35 mm	12.7 mm	25.4 mm	50.8 mm	76.2 mm	101.6 mm
	1/4"	6.35 mm	12.7 mm	25.4 mm	50.8 mm	101.6 mm	152.4 mm	203.2 mm
	1/2"	12.7 mm	25.4 mm	50.8 mm	101.6 mm	203.2 mm	304.8 mm	406.4 mm

Figure 4.39 Overall lay-up width for common lay-up technologies and configurations

4.4.4.2 Cost Effectiveness of Automated Lay-up Processes

Floor-to-Floor Productivity

The productivity of the process is not determined solely by the lay-up productivity mentioned above. Additional operations that affect machine availability contribute significantly to the effective overall productivity. Floor-to-floor productivity considers both the lay-up productivity as well as the productivity decrease caused by processes that affect available machine time, such as the following:

1. *Lay-up preparation:*
   - Mounting of mold, and where applicable preparation of lay-up equipment
   - Measurement and calibration/tooling
   - Preparation of mold surface on the machine envelope
   - Machine and process tests, dry runs, etc.
2. *Non-productive time intervals during lay-up:*
   - Material loading
   - Cleaning and maintenance operations during the lay-up phase
   - Quality inspection
3. *Post-processing:*
   - Demolding of component or other works within the machine envelope (bagging, etc.)
   - Unloading of mold

Depending on machine configuration and specific application, these time periods may play an important role in comparison to the perfect lay-up time. Therefore, accurate adaption of lay-up technology and additional processes to the intended application is crucial.

Additional Productivity Criteria

In addition to machine productivity (e.g., floor-to-floor productivity) other criteria need to be considered carefully in order to evaluate the cost effectiveness of a technology. Notably considerable are recurring costs as well as costs of capital investment and non-recurring costs.

1. *Recurring costs:*
   - Material costs (tape material, auxiliary materials, consumables, spare parts, software licenses, etc.)
   - Personnel costs (machine operation, maintenance, programming, etc.)
   - Building costs
   - Cost of capital investment and interest
2. *Capital investment and non-recurring costs:*
   - CAD model preparation
   - Programming of components
   - Lay-up tests for program optimization (material and machine utilization, etc.)
3. *Utilization of the lay-up equipment for additional operations and processes:*
   - Trimming, lay-up of additional materials, such as glass, copper mesh, etc.
   - Evaluation of savings in other equipment and process time
   - Downtime of equipment because of additional operations
4. *Machine downtime caused by maintenance:*
   - Scheduled maintenance and cleaning
   - Non-scheduled repairs

Unit Price per Component

In the end, all considerations regarding productivity can be translated into a unit component price. In general, the component price will be the criterion to determine the economic viability of a lay-up machine and lay-up configuration. Such an analysis is also expedient in order to relate capital investment and structural costs to the effective productivity. The obvious goal is the selection of the optimum machine configuration for a specific component spectrum.

4.5 Pultrusion

The pultrusion process is one of the longest standing processes for the manufacture of continuous fiber reinforced plastic components, and at the same time it is the oldest continuous process. In a typical pultrusion process (Figure 4.40), reinforcing fibers are impregnated with a matrix (thermoplastic or thermosetting) using a wet-out process and subsequently cured in a forming die. Here, the profile is continuously pulled through the die and can subsequently be cut to length with a saw. Today, this process facilitates the automated and cost-effective manufacture of fiber-reinforced plastic profiles of various geometries. The name pultrusion is derived from "to pull" and "extrusion". Depending on the mechanical characteristics that the profile to be manufactured should exhibit (e.g., stiffness), a broad range of different fiber textiles can be utilized, such as rovings, randomly oriented fiber mats, wovens, stacks, or 3-D fabrics. The process can be combined with others, e.g., with a braiding process, also called "pull-braiding", which is particularly suitable for round cross sections.

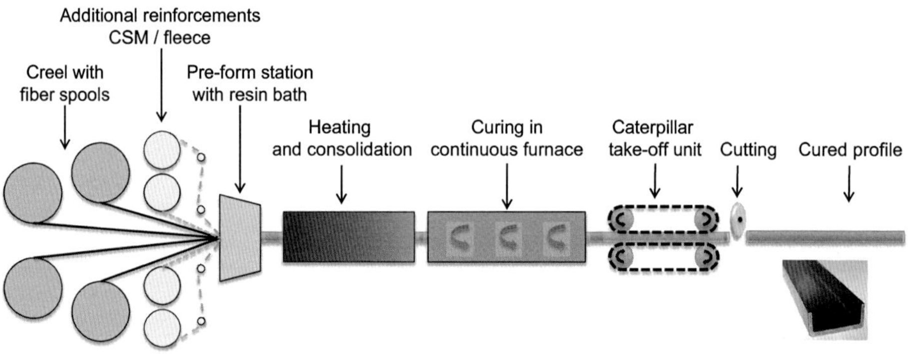

Figure 4.40 Pultrusion line, schematic

The pultrusion process can be differentiated in two process variations. During the course of an *open process*, the reinforcing fibers are pulled off doffs or rolls from a creel racking system and guided over a dip roller through a resin bath. A carding index provides the desired fiber distribution/orientation in the manufactured profiles. The fibers are wetted out with resin in the resin bath and may pass through one or more pre-shaping stations that increasingly consolidate the resin impregnated fibers and slowly approximate the final, desired profile. In the course of a *closed process*, the reinforcing fibers are exposed to the resin only when entering the shaping die. Here, the liquid resin is pushed through the die under pressure. This approach allows for higher fiber volume contents than those achievable in open processes. In addition, this process allows for a significant reduction in emissions, e. g., from styrene-containing resins (UP, VE). In the curing die, which is typically heated by electric heaters, the impregnated material is quickly brought to curing temperature. Depending on resin system and type (EP, UP), curing temperatures range between 100 and 180 °C. Usually, the profile slowly heats up from the outside to the inside (depending on die configuration), causing the resin to gel and cure. Simultaneously, friction between the component and the die surface increases drastically. In order to reduce this friction, internal release agents can be added to the resin system, or the die surfaces can be coated with suitable release agents. For high-volume profiles it is of particular importance to maintain constant heat distribution in order to prevent cracks in the matrix that may be caused by different degrees of curing between the outer and inner layers of the laminate. The cured profile is subsequently cut to desired lengths. The pultrusion process is driven by a gripper/puller system, for example in the form of a reciprocating hydraulic clamp system or an opposed tread caterpillar-tractor type system.

Besides these two process variations, the direct pultrusion processing of prepregs is gaining increasing application. The advantages of using a semi-finished material again include the fact that the ratio of fiber to matrix is already optimized and that high fiber volume contents (60 to 65 %) can be achieved. Prepreg materials also allow for the use of highly viscous matrix resins (e. g., by using tougheners), which cannot be utilized in open resin bath processes. Therefore, even high-performance prepregs used in aviation applications can be pultruded. Another advantage of these systems is the fact that they do not release solvents or styrene.

Figure 4.41 depicts the process sequence. In contrast to the two pultrusion variations described above, there is no external resin supply when prepregs are pultruded. Instead, prepregs are pulled from reels, pre-shaped by guide pulleys, and then pressed, heated, and cured in an oven section. Depending on geometry and utilized prepreg material, the prepregs are cut into narrower tapes before being fed into the pultrusion line. Here, the prepregs are first unwound, then cut to the required width that is determined by its position in the laminate structure, and finally rewound. This approach allows for the manufacture of complicated layer structures with varying orientations, as well with combinations of UD tape and fabric prepregs, such as stringers or ribs required in the aircraft industry. For example, JAMCO, Japan, uses prepregs to manufacture continuous stringer profiles for airplanes in the Airbus A320 series and the A380. Here, double-T stringers are manufactured from flat prepreg cuts, as shown schematically in Figure 4.41. Several layers of UD tape and fabric prepreg are pre-assembled, and then the strips are fed through idler and pre-form pulleys into the pultrusion system, intermittently shaped, and cured.

Similar to traditional pultrusion processes, the friction between component and die surfaces increases along the die. The die surfaces have to be coated with suitable release agents, because it is not feasible to add these agents to prepreg systems in order to reduce friction at the die surface. The cured profile is subsequently cut to length. This pultrusion process is also driven by a caterpillar-tractor type system. For prepreg systems typical in the aviation industry, the material feed may be clocked to facilitate the longer curing time so that in fact this process should be considered quasi-continuous.

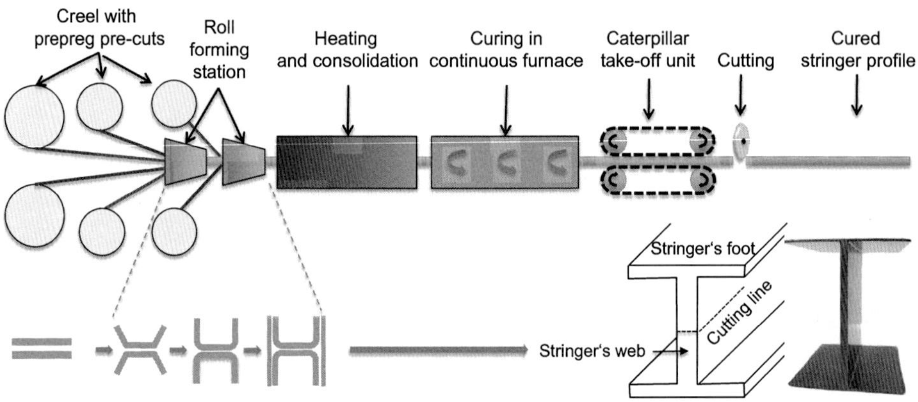

Figure 4.41 Pultrusion process for prepreg materials

Secar Technologie GmbH, Austria, manufactures sheet, angle, and hollow profiles for various industrial and aviation applications from fast-curing prepreg systems (with curing temperatures ranging from 80 to 150 °C) using a fully continuous pultrusion process (Figure 4.42).

Figure 4.42 Pultrusion profiles [Courtesy: Secar Technologie GmbH]

Profiles/Gusset Filler

Today, the technology has reached a level of maturity that allows for the manufacture of not only straight, but also of curved profiles with a defined radius. Curved profiles, as well as profiles with varying cross sections, are of particular interest in the aviation industry because they facilitate lightweight design using fiber reinforced materials. Currently, profiles such as spars and stringers, are manufactured using dies specifically adapted to each component. The corresponding prepreg stacks are laid up, trimmed, pre-shaped via hot forming, and finally cured in a tooling device. Here, the utilization of adapted prepreg pultrusion may lead to a significant simplification and cost reduction.

Besides cured profiles, it is also possible to manufacture triangular gusset fillers from pre-impregnated materials, which is of particular interest in the aviation industry. The filler shown in Figure 4.43 is used to cover the fillet weld that is created when two L-shaped profiles are joined to form a T-profile (stringer). The fillet weld would cause porosities and changes in fiber orientation during curing of the stringer profile. Therefore, it is often filled with the same material (Figure 4.43).

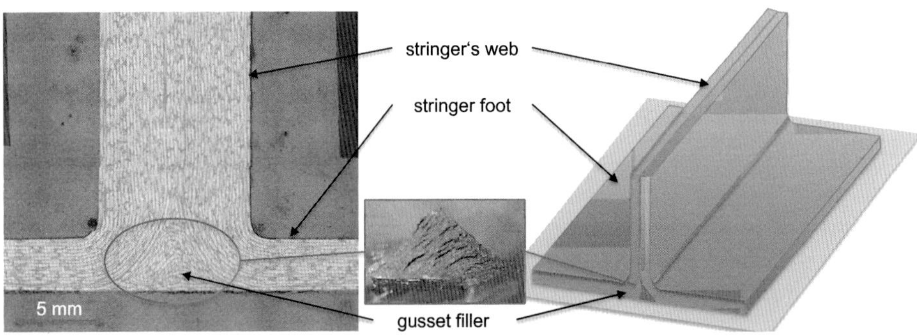

Figure 4.43 Gusset filler in the foot of a T-stringer

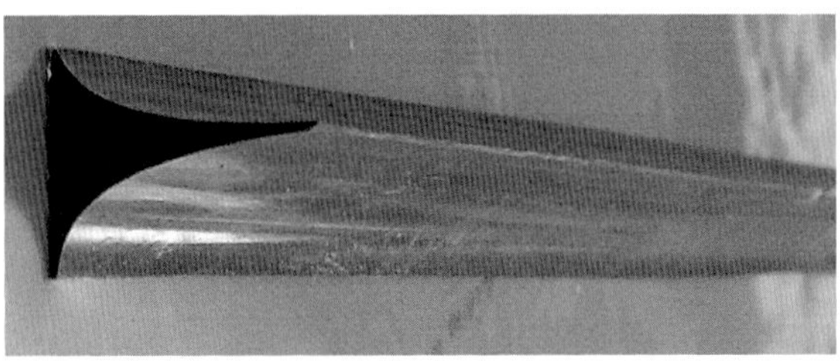

Figure 4.44 Gusset filler made from prepreg, uncured [Courtesy: Secar Technologie GmbH]

4.6 Hot Drape Forming

Flexible, adaptable processes and tools are required in order to produce the wide variety of often complex composite components on an industrial scale from flat, pre-impregnated prepreg stacks.

Shaping technologies based on hot pressing methods have been established (Figure 4.45); however, they are not flexible and are very limited in terms of their geometrical dimensions because of their massive tools. Another significant disadvantage of pressing technologies is the fact that they do not allow for easy realization of complex geometries, such as undercuts. Often, the press is used for both the shaping process of the materials as well as the subsequent curing process. In these cases, very short curing cycles are required in order to facilitate the next shaping cycle. Fiber reinforced materials also require close attention with respect to their typical anisotropic behavior, which results in directional dependence of material properties. In light of the fact that continuous carbon or glass fibers inhibit the flow and elongation of the material, the use of hot pressing methods for large-volume, industrial shaping applications of flat prepreg stacks is limited, and requires specifically adapted concepts and materials.

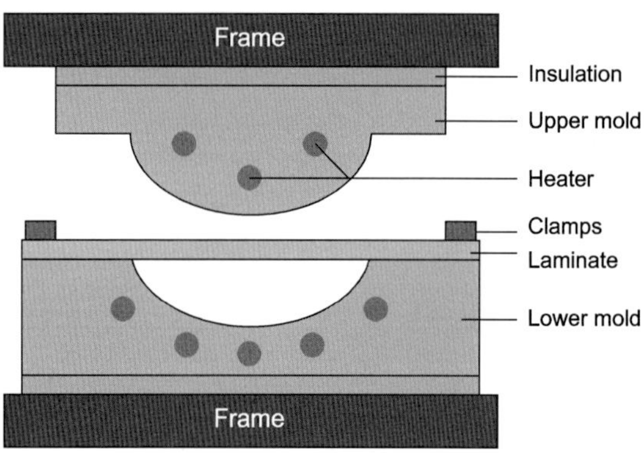

Figure 4.45
Hot pressing technology [4]

Therefore, so-called "hot forming" technology is often used to process and shape high-performance prepregs. The concept of hot forming is derived from thermoforming processes used for thermoplastic materials. It facilitates the manufacture of often highly complex geometries, such as multi-curved stringers or ribs/fins from flat, two-dimensional prepreg stacks (produced by ATL and automated cutting systems) that can subsequently be integrated into the final components in their uncured (B-stage) state. Examples from the production of aviation components include the wing sheeting of the Airbus A350XWB, for which stringer elements as long as 33 m manufactured by hot forming are integrated with the pre-cured sheeting skin, or the C-spars of the vertical tail of the A380, for which flat, up to 10 mm thick and 6 m long prepreg stacks are formed and assembled into C-profiles (Figure 4.46).

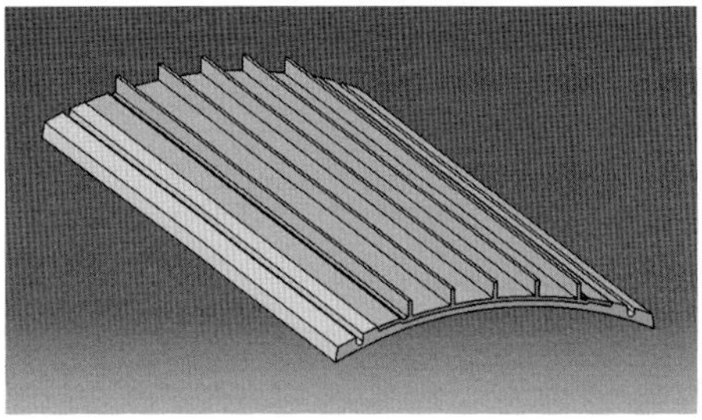

Figure 4.46
Panel reinforced by stringers [Courtesy: Ina Steinicke]

Hot forming utilizes diaphragm forming equipment that is derived from thermoforming of thermoplastic materials. We distinguish between single- (SD) and double-diaphragm (DD) equipment (Figure 4.47 and Figure 4.48). The prepreg stack is heated in the diaphragm machine by either IR heaters, hot air blowers, or inductive heaters (1). Next, the material (3) is formed on the tool (4) with the help of flexible rubber membranes (diaphragms) (2) that are stretched over tentering frames. Vacuum is created between membrane and tool, which slowly presses the membrane and thus the prepreg stack to the tool. Typical forming temperatures for pre-impregnated materials (prepregs or fabrics/stacks with binders) range between 60 and 80 °C. The cycle times for hot drape forming (heating, forming, cooling) vary considerably, depending on laminate thickness or number of plies. Both variations are suitable for almost any geometry and offer very good forming results even for parts with undercuts.

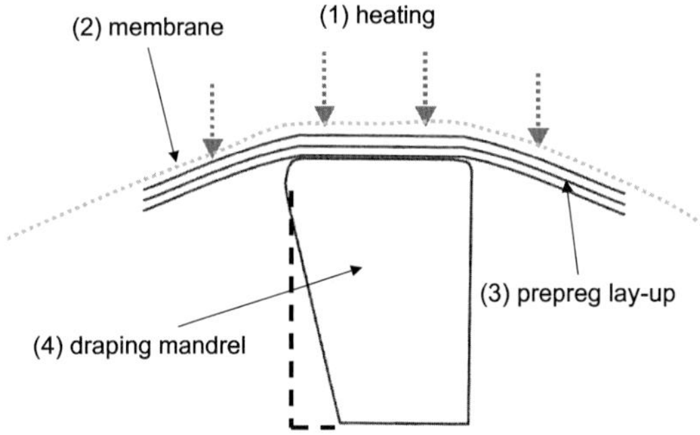

Figure 4.47 Single-diaphragm process (SD process) [4]

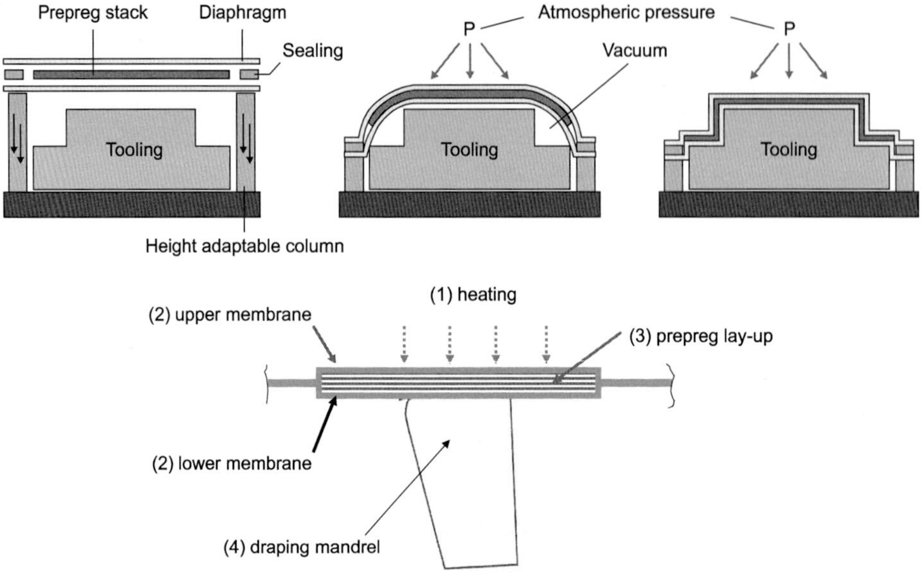

Figure 4.48 Double-diaphragm process (DD process) [4]

For high-quality forming operations it is critical to allow for individual plies sliding over each other (interply slip) during the forming process that transforms a flat into a curved geometry. The ply slip in prepreg stacks is often very limited because of the typically high matrix viscosities at room temperature. Therefore, the material must be uniformly heated in order to reduce matrix viscosity and allow for maximum interply slip. After forming, successful ply slip can be identified by the so-called "book effect" (Figure 4.49). However, care must be taken when heating the material in order to prevent premature curing of the matrix. The B-stage of the prepreg matrix is also of importance, because it limits excessive resin flow and thus "bleeding" of the prepregs when heated.

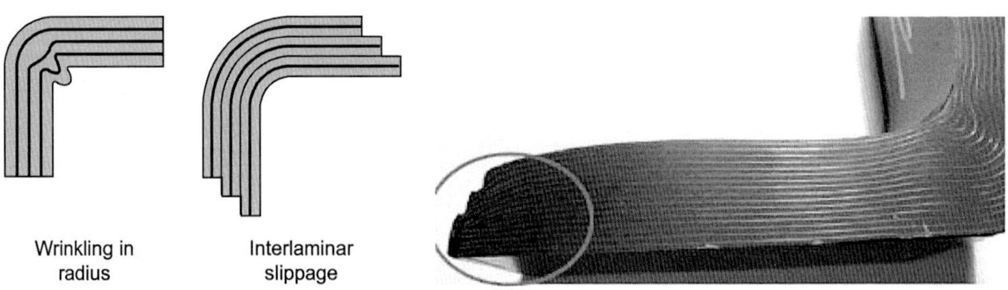

Figure 4.49 Formed laminate without (left) and with "book effect" [4, 5]

Insufficient slip generally results in disturbed fiber courses, which in turn cause wrinkles/buckles and void formation in the laminate, reducing the mechanical properties of the composite (Figure 4.50).

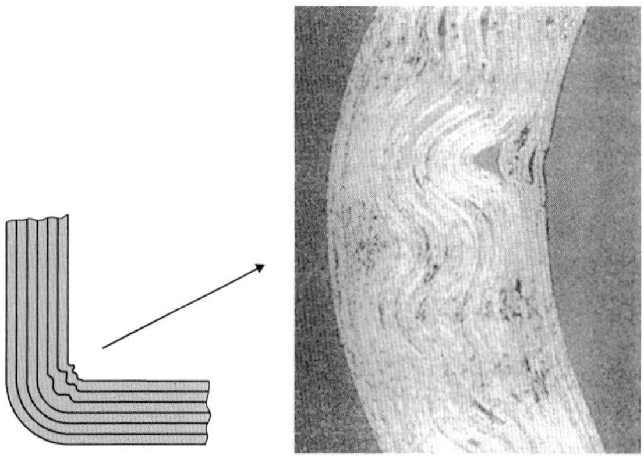

Figure 4.50
Voids and fiber buckles in the interior of the laminate [4, 5]

Essential quality criteria for successful forming operations include the following:

- Slipping of individual plies over each other (interply slip)
- Length differences along laminate edges parallel to the forming line
- Edge formation = book effect

In preparation of SD processes, a flat, trimmed prepreg stack is positioned on a forming tool (made from metal or hardwood). Thermal radiation from ceramic or IR heaters, or from electric bulbs, heats the membrane from above. Thermal conduction will then increase the temperature throughout the laminate thickness. The temperature increase on the membrane's top face typically follows a linear profile (or ramp) until it reaches a set temperature. Once this temperature is reached, an integrated controller maintains the temperature on the membrane top face while the stack continues to heat and approaches the membrane temperature. There is always a temperature gradient between membrane top face and laminate bottom face because of cooling effects below the membrane and the insulating effect of the diaphragm. Metal forming tools in particular may allow for uncontrolled heat loss from the stack into the tool. This thermal flow creates an additional temperature gradient within the stack and may cause the formation of wrinkles/buckles in the laminate, as discussed earlier (Figure 4.50).

It is not mandatory to heat the entire stack width, because only the so-called "slipping width" in the forming area requires heating. Therefore, heating systems have been developed that allow for heating from below, directly at the stack, via hot air (Figure 4.51).

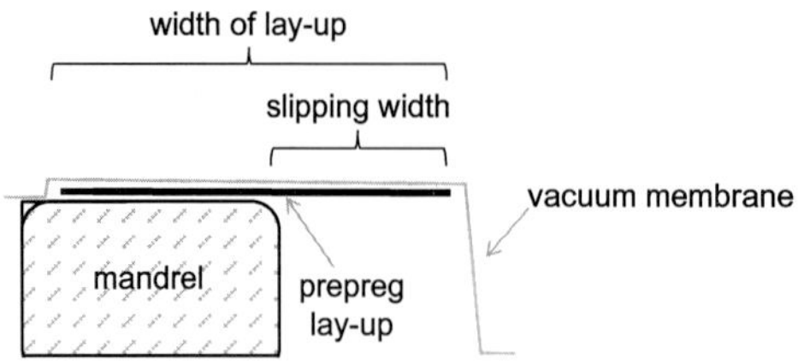

Figure 4.51
Heating of "slip width" of a prepreg stack

The DD process is slightly more complex in terms of process chain and machine configuration than the SD process, because the prepreg stack has to be positioned between two membranes before it can be placed on the tool to be formed. However, the forming results are of even higher quality than those achieved by SD processes, in particular when thick prepreg stacks (up to 10 mm) are formed. The stack can be heated from both sides and heat loss into the forming tool can be avoided.

Commercial hot drape forming systems are offered by Premium Aerotec Varel, Germany or Laminating Technology, UK, among others.

Figure 4.52 Hot drape former [Courtesy: Laminating Technology]

Independent of the chosen forming technology it is important to consider that continuous fiber reinforced materials require component geometries that do not restrict interlaminar fiber slippage, because in contrast to short-fiber reinforced materials and metals, prepreg materials cannot be plastically deformed in the fiber direction.

In the following we will describe a possible approach using a swaged hand-hole geometry (Figure 4.53) as an example.

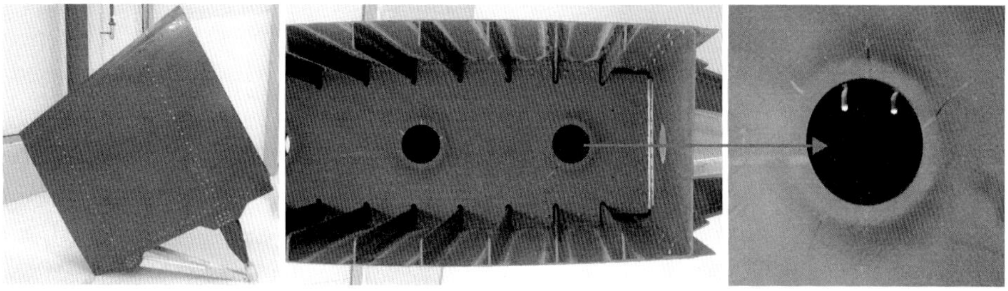

Figure 4.53 Example of a vertical tail plane made from carbon fiber reinforced plastics, ribs/fins with swaged hand-holes

In order to provide enough range of movement for the fibers in the boundary area of the reinforcing ribs/fins, a circular hole has to be cut in the flat prepreg stack. Also, the tool geometry has to be adapted to both the selected process and material. The radii of the ribs/fins should be considered in particular (Figure 4.54). Note that the shear stress gradient at the base of the rib/fin increases with decreasing lower radius R_l which in turn increases the effect of fiber elongation (Figure 4.54 und Figure 4.55). At the same time, R_l influences the ramp length *l*. Thus, with increasing ramp length *l*, the shear stress gradient decreases, and therefore the fiber length difference also decreases. As a result, the risk of wrinkle/buckle formation in fiber orientation will decrease.

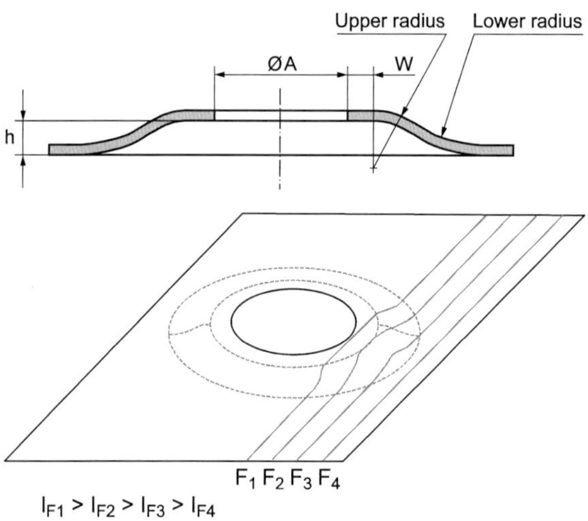

Figure 4.54
Example geometry: reinforcing ribs/fins with swaged hand hole [4]

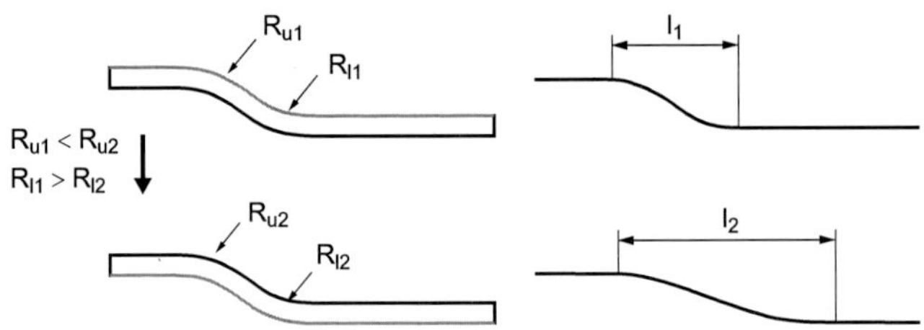

Figure 4.55 Effect of varying radii (R_l = lower radius, R_u = upper radius) [4]

For the realization of a rib/fin reinforced hand hole made from DU prepregs (quasi-isotropic laminate structure), the appropriate forming radius can be determined, e.g., by preliminary tests so that the requirements mentioned above as well as the static requirements for the geometry are met. In light of the fact that the forming temperatures are relatively low (60 to 80 °C) and do not require a metal tool, hardwood may be chosen as the tool material. This will also keep the expenditure necessary for the construction low and minimizes heat loss from the stack to the tool. A challenge of DD processes is the required accuracy with which the stack and the membranes have to be positioned on the tool. In our example, the tool was a designed like a thermoforming tool that provides an elevated edge for external line-up and positioning of the laminate.

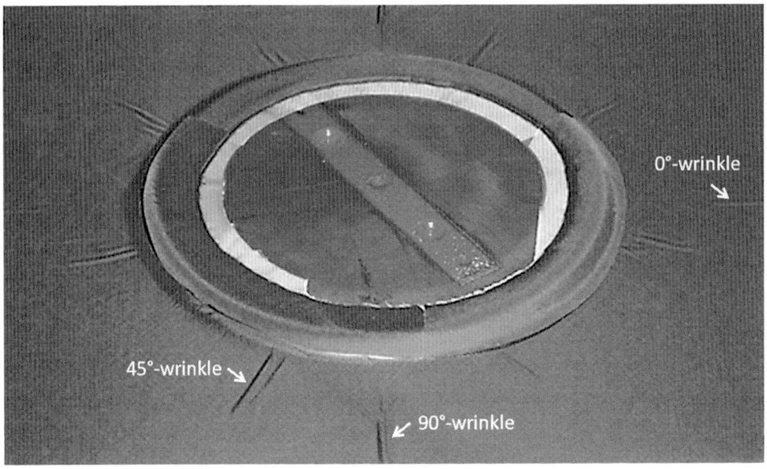

Figure 4.56 Forming tool for double-diaphragm process [4]

Figure 4.56 shows a swaged hand hole manufactured by hot drape forming (DD) process. Slight wrinkles/buckles in the 45° direction can still be detected. Although a hole was cut in the flat stack and the radii were adapted to component and tool, it was not possible to completely prevent the formation of these wrinkles/

buckles because of the differences in fiber length. However, the example demonstrates that it is feasible to hot-form UD prepreg stacks into a 3-D geometry.

4.7 Same-Qualified-Resin-Transfer-Molding Processes (SQRTM)

Substitution of the conventional autoclave curing process with out-of-autoclave (OoA) processes provides the potential to increase the economic viability of the manufacture of larger components. Over the last several years, a number of these OoA processes have been introduced on the market. However, only few of these can compete with same-qualified-resin-transfer-molding processes. The resin transfer molding (RTM) was modified and further developed under the acronym SQRTM so that large, complex, near net shape components, e. g., helicopter cabin roofs, landing flaps, or floor cross beams, can be manufactured. Here, prepreg materials are processed together with liquid resin utilizing infusion technology. This RTM process for UD and fabric prepregs was developed by Radius Engineering Inc., Salt Lake City/Utah, and is currently employed by several aviation component manufacturers, who either use it in modified versions or further investigate its implementation [6–8]. This technology produces high-quality, near net shape components for the aviation industry, without the use of autoclaves.

The fundamental difference between SQRTM and traditional resin injection processes (RTM) is the lay-up of UD or fabric prepregs or prepreg cuts instead of the manufacture of a so-called preform from dry fabrics.

Here, the cuts are draped in an RTM tool and pre-consolidated when the tool is closed (typically under vacuum). Then the infusion resin is pumped into the RTM tool under high pressure (approx. 5 to 7 bar), thus exerting hydrostatic pressure on the prepreg stack. The purpose of the injected resin is not to wet the prepreg, but rather to uniformly apply pressure to achieve consolidation (comparable to autoclave curing). Here, the infusion resin may be considered an auxiliary material and under certain conditions may not have to be qualified again, as is otherwise mandatory for materials in aviation applications.

Both materials are cured together, a process during which the infusion resin creates a uniform, smooth surface on the prepreg stack. This technology combines the advantages of prepreg materials, which allow for a high degree of lightweight construction, with those of infusion processes.

Advantages of SQRTM:

- It allows the use of impact-modified and other special resins
- Prepregs are easier to handle than dry fabrics
- An established and proven layer structure can be maintained with prepregs
- Established and compatible prepreg and infusion resin systems can be utilized; no additional material qualification process required
- Process and process parameters (pressure, temperature), component thickness, and fiber volume content are easier to control
- The use of heated presses facilitates shorter curing times (faster heating and cooling rates)
- Out-of-autoclave curing is possible

An additional advantage is the fact that larger structures, such as landing flaps or horizontal stabilizers, can be manufactured as integral components (multi-spar concept), which significantly reduces assembly effort. Because such components have to be manufactured in a closed tool, they exhibit their final dimensions (net shape) after demolding, so any machining operations are limited to very few areas (deflashing). Therefore, it is possible to achieve levels of dimensional accuracy with SQRTM processes that are difficult to achieve with prepreg components manufactured by autoclave curing.

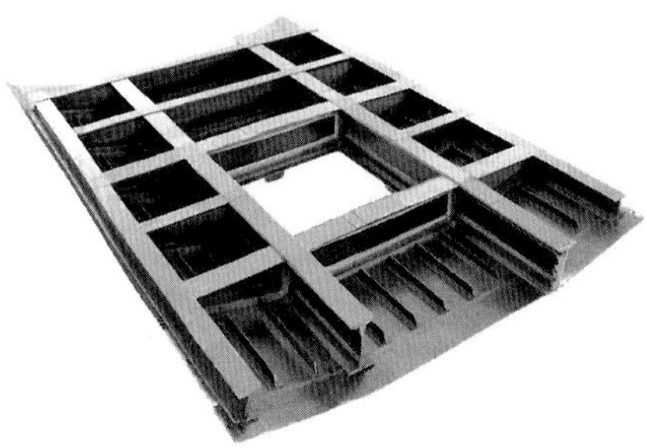

Figure 4.57
Near net shape helicopter roof element [Courtesy: Radius Engineering Inc.; Salt Lake City [8]

4.8 References

[1] Patent EP 1775108 B1: Martinez Manuel Torres. *Multi-Application Head for Fibre Strips and Method of Applying Said Fibre Strips*. 2011.

[2] Patent US 20110088527 A1: Martinez Manuel Torres. *Strip-Cutting System*. 2011.

[3] Patent US 20080216961 A1: Alexander Hamlyn, Yvan Hardy. *Applicator Head for Fibers with Systems of Cutting and Locking Particular Fibers*. 2008.

[4] Hauke Lengsfeld, Christopher Werner. *Warmumformen statt Pressen*. Kunststoffe. 2008, 3, 28–31.

[5] Charles L. Tucker III. *Forming of Advanced Composites*, in Advanced Composites Manufacturing. Timothy G. Gutowski (Ed.). New York: John Wiley & Sons Inc., 1997.

[6] Sara Black, Jürgen Schreier. *SQRTM ermöglicht die Fertigung endkonturnaher Teile aus Verbundwerkstoffen*. Maschinenmarkt. (Online) http://www.maschinenmarkt.vogel.de/themenkanaele/produktion/kunststoffverarbeitung_gummiverarbeitung/articles/293002/index5.html (accessed 5/16/2012).

[7] Cederic de Roover, Bertrand Vaneghem. *Highly Integrated Structure Manufactured in One-Shot with Prepreg UD Tape*. JEC Composites Magazine. 2011, 62, 40–42.

[8] Sarah Black. *SQRTM Enables Net-Shape Parts*. High Performance Composites. September 2010.

4.9 Additional Information

- M. Flemming, G. Ziegmann, S. Roth. *Faserverbundbauweisen – Fertigungsverfahren mit duroplastischer Matrix*. Berlin, Heidelberg: Springer Verlag, 1999. ISBN 3-540-61659-4.
- M. Flemming, G. Ziegmann, S. Roth. *Faserverbundbauweisen – Halbzeuge und Bauweisen*. Berlin, Heidelberg: Springer Verlag, 1996. ISBN 3-540-60616-5.
- Company brochure: assyst Bullmer GmbH und Gunnar/Weissenberger AG, 2012.
- J. Sloan. *High Performance Composites*. Composites World. July 2008.
- Frank Henning, Elvira Moeller. *Handbuch Leichtbau: Methoden, Werkstoffe, Fertigung*. Munich: Carl Hanser Verlag, 2011. ISBN: 978-3-446-42267-4.
- Patent EP 2305455 A1: Martinez Manuel Torres. *Cutting system for Strips*. 2011.
- Patent US 20130111725 A1: Martinez Manuel Torres. *Application Head for Applying Fiber Strips*. 2013.
- Patent EP 2036702 B1: Alexander Hamlyn, Yvan Hardy. Fiber Application Machine.
- Patent EP 2117818 B1: Alexander Hamlyn, Yvan Hardy. *Faserapplikationsmaschine mit flexiblen Faserzufuhrschläuchen*.
- Patent EP 0361828 B1: Jerry Dale Wisbey. *Fiber Placement Machine, and Fiber Placement Head*. Cincinnati Milacron Inc.
- Patent US 8151854 B2: Tino Oldani. *Fiber Placement Machine Platform System Having Interchangeable Head and Creel Assemblies*. Ingersoll Machine Tools, Inc.
- Patent EP 1922209 B1: Russel C. Devlieg, Kenneth P. Zaballos. *Tow Cutting Device, System and Method*. The Boeing Company.
- Patent US 8256484 B2: Robert A. Kisch, Peter Vogeli, Kyle Jeffries, Russell DeVlieg. *End Effector for Constructing Composite Members*. The Boeing Company.

- Patent WO 2008154269 A8: Milo M. Vaniglia. *Motorized Cut and Feed Head.* Cincinnati Machine, Llc. 2008.
- Patent US 5110395 A: Milo M. Vaniglia. *Fiber Placement Head.* Cincinnati Milacron Inc.
- Russell DeVlieg, Kyle Jeffries, Peter Vogeli. *High-Speed Fiber Placement on Large Complex Structures.* Electroimpact, Inc., Copyright © 2007 SAE International. Paper No. 2007-01-3843.
- Luis Izco, Javier Isturiz, Manu Motilva. *High Speed Tow Placement System for Complex Surfaces with Cut/Clamp & Restart Capabilities at 85 m/min (3350 IPM).* 2006 SAE International, Paper No. 2006-01-3138.
- Dirk H.-J. A. Lukaszewicz, Carwyn Ward, Kevin D. Potter. *The Engineering Aspects of Automated Prepreg Layup: History, Present and Future.* Composites Part B: Engineering. 2012, 43, 997–1009.
- Fives Machining Systems, Inc. www.fivesmsi.com (accessed 06/12/2014).
- Michael Chun-Yung Niu. *Composite Airframe Structures – Practical Design Information and Data.* Hong Kong: Hong Kong Conmilit Press LTD., 1992.
- PCT Patent WO 2006/069989: *Process for Making Swaged Lightening Holes in Planar Areas of Preimpregnated Composite Parts.* 2006.
- Dave Regiec, Peter Vogli. *Highspeed Fiber Placement on Large Complex Structures.* JEC Composites Magazine. 2012, 72, 41–43.

5 Curing Technologies

Felipe Wolff-Fabris

Chapter 4 introduced suitable technologies to form prepreg materials. Once forming operations are completed, they are followed by the curing process, which utilizes either autoclaves or out-of-autoclave processes. The curing of a prepreg component is the result of chemical reactions that lead to the final cross-linking of the matrix system. The energy required to initiate chemical cross-linking may be introduced in a variety of ways, including heat, electromagnetic waves, or electron beam radiation.

In this chapter, we will introduce the main technologies commonly used to cure prepregs. The advantages and disadvantages of individual processes and the effect of their respective process parameters on final component characteristics will be discussed. The following section will cover the fundamentals of the curing process, in particular those crucial for the adaptation and optimization of the process parameters.

5.1 Fundamentals of the Curing Cycle

The chemical reaction of common resin system have already been described in Chapter 2. Conventional, thermally curing systems (e.g., epoxy-amine) can basically be cured utilizing any of the technologies introduced here. Only electron beam curing represents a special case and requires special resin systems. The chemical reactions and curing cycles of electron beam curing prepregs will be discussed separately in Section 5.8. In the following we will discuss the fundamental aspects of the curing cycle for common thermally cured prepregs as they relate to the technologies detailed later.

Prepregs are typically transported at −18 °C and can be stored at this temperature for months and even years before they begin to cure. Even at room temperature prepregs react relatively slowly; here, the pot life may reach several weeks. The term "pot life" describes the time span until the system is no longer processible because of its increased viscosity. In order to completely cure prepreg materials, high temperatures have to be maintained for a certain period of time. Prepreg components for the aircraft industry are typically cured at 180 °C for several hours. The high temperatures are necessary to initiate and maintain the chemical reaction. Ideally, all reactive groups will have formed a cross-linking point after the curing cycle is completed. In this case, post-curing and the resulting dramatic changes in characteristics over the component's service life can be excluded.

The chronological sequence of temperature, and pressure where applicable, required for the curing of a component is also called *curing cycle*. Considering the fact that the productivity of prepreg manufacturing processes is largely dependent on the curing cycle, the latter should be as short as possible. The curing cycle is affected by several aspects, including resin chemistry, use of accelerators, and curing temperature. The chemical aspects of the curing reaction, depending on utilized resins and curing agents, as well as the change in viscosity during cross-linking, have already been extensively discussed in Chapter 2.

Figure 5.1 shows a typical curing cycle for prepregs in the aircraft industry. It involves two heating ramps and two isothermals. During the first heating ramp, the resin viscosity decreases significantly and cross-linking reactions do not yet occur. Subsequently, the temperature is maintained at approx. 115 to 140 °C. In this isothermal region the resin is able to flow and volatiles will diffuse out. This isothermal is also important to facilitate the homogeneous heating of the prepreg through the entire laminate thickness. In response to the additional thermal energy of the second heating ramp the viscosity decreases even further until the cross-linking reaction of the matrix is initiated, rapidly increasing the resin viscosity. The curing process continues during the second isothermal. The liquid resin develops a gel-like structure; this transition is also called gelling or gel point. The second isothermal should be sufficiently long to facilitate the reaction of almost all reactive groups into cross-linking points.

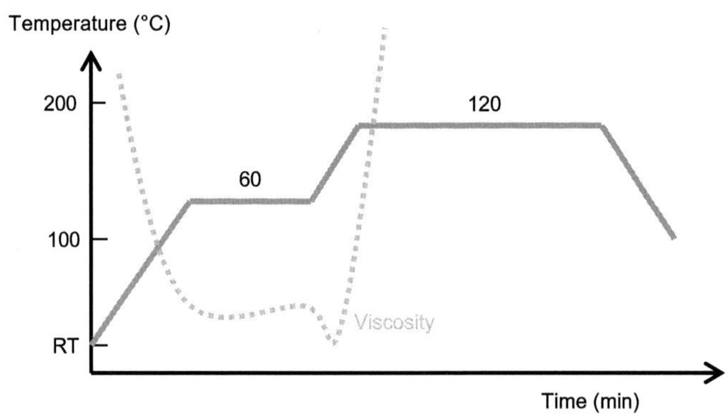

Figure 5.1 Typical curing cycle of prepreg components

The chemical curing reactions of resins are exothermic (see also Chapter 2). This means that heat is released during curing, which has to be taken into account during the heating phase. If the heating ramp is too steep, the large quantity of heat released in a short time leads to high temperature increase in the laminate. On the one hand, this may lead to a non-uniform heat distribution in the component and thus to an inhomogeneous cross-linking density (= inhomogeneous properties). On the other hand, the decomposition temperature of the matrix may be exceeded, which would lead to a significant increase in porosity and thus to a dramatic loss in mechanical properties. The laminate may even burn, which would not only have disastrous economic consequences, but poses an accident hazard. Therefore, the heating ramp has to be adapted to the prepregs utilized as well as to the component thickness.

During the first isothermal of a curing cycle, excessive resin is absorbed by a so-called "bleeding aid". Bleeding during the curing of a prepreg describes the flow of excessive resin from the component. This process can be supported by external pressure. Resin flow during the curing cycle is of particular importance because it facilitates the removal of entrapped air and volatiles from the component and the reduction of laminate porosity. The fraction of resin bleeding from the laminate depends on time and resin viscosity, which in turn depends on the heating ramp (Figure 5.2). Lower viscosities result in higher resin flow, which can only occur prior to reaching the gel point. Reducing the temperature of the first isothermal delays resin gelation (Figure 5.3), thus extending the duration of resin flow.

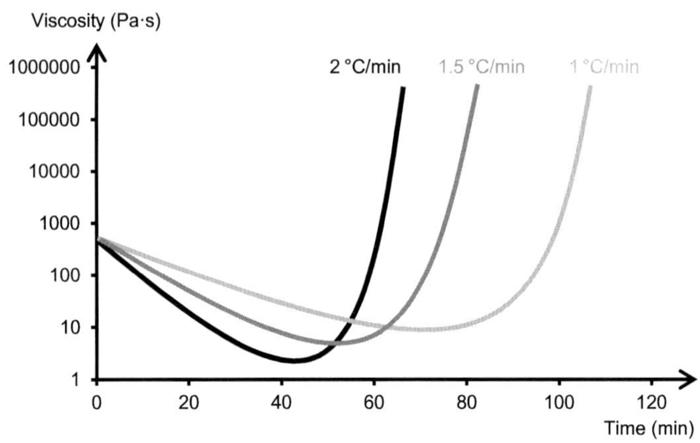

Figure 5.2 Effect of heating ramp on resin viscosity

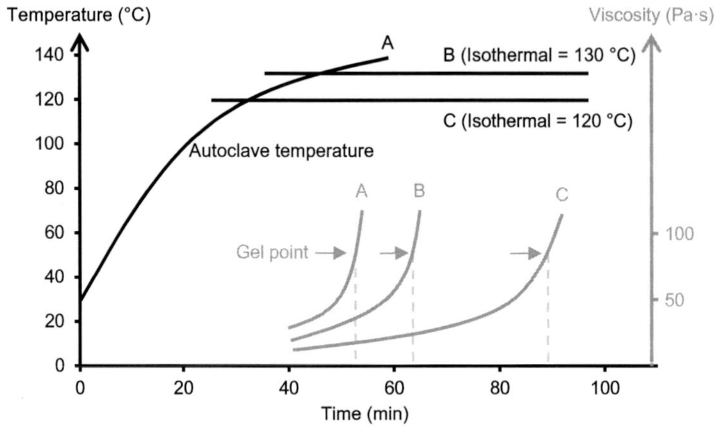

Figure 5.3 Effect of isothermals on gel point [1]

Over the last several years, "low bleeding" prepregs have been introduced to the market. Here, bleeding occurs mainly during the first isothermal. The resin content of these new prepreg systems (so-called zero-bleed prepregs) is precisely adjusted to match the desired fiber volume content and layer thickness in the final component. Thus, it is possible to shorten, or even eliminate, the first isothermal, as long as the diffusion of volatiles is ensured.

Typically, a vacuum bag design is used to cure prepregs using autoclave, oven, Quickstep™, microwave, or electron beam technologies. The vacuum ensures that air is removed from the prepreg in order to reduce defects (porosities and delamination) as much as possible. In addition, the vacuum bag will retain excess resin flowing from the laminate. The utilization of vacuum bags facilitates an increase in component quality and an improvement in their final characteristics. Figure 5.4 shows a prepreg laminate in a vacuum bag, as well as the schematic design of a vacuum bag.

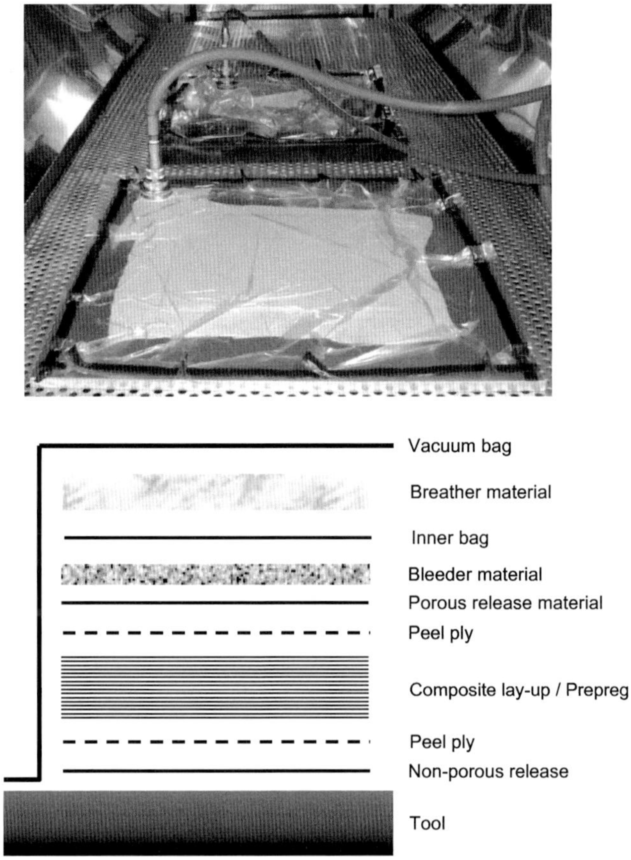

Figure 5.4 Typical vacuum bag design [Courtesy: SKZ – Das Kunststoff-Zentrum]

■ 5.2 Autoclave Curing

The most common method of prepreg curing is the use of autoclaves. They are of particular advantage for the manufacture of components with varying geometries, because the pressure is built up isostatically in the chamber. Thus, the component surfaces are exposed to identical conditions, independent of their size and geometry. Autoclave size limits the maximum component size that can be manufactured using this technology. For example, an autoclave with a length of 23 m and a diameter of 9 m is used for the production of fuselage sections of the Boeing 787 Dreamliner.

A typical autoclave consists of a pressure chamber, electrical and digital control systems, a hot gas generator, and a vacuum system (Figure 5.5). An autoclave can accommodate and cure individual large or several small components. Therefore, autoclaves exhibit large utilization flexibility.

Figure 5.5
Autoclave for curing of small components [Courtesy: SKZ – Das Kunststoff-Zentrum]

The following requirements need to be considered when selecting curing temperature and pressure:

1. The resin cures homogeneously and reaches the desired degree of cross-linking in as short a time as possible.
2. The maximum allowable material temperature (= decomposition temperature) must not be exceeded at any point of the component.
3. The curing pressure is sufficiently high to allow for the flow of excessive resin from all prepreg layers.

Commonly, the pressure inside the autoclave is created by an inert gas such as N_2 or CO_2. Air can also be used; however, it increases the fire hazard at high curing temperatures. In general, pressures up to 7 bar are used to consolidate the prepregs and to reduce porosity. In addition, the desired temperature in the component is reached faster at higher pressure, because convective heat transfer becomes more effective at higher gas pressures and velocities and stronger gas turbulence. Pressure also creates shear forces that increase resin flow and fiber wetting. All these factors affect improvements in mechanical properties of the final component. The pressure transfer into the component depends on several parameters, including fiber volume content, laminate configuration, and design of the vacuum bag.

Figure 5.6 shows a conventional pressure profile during the curing cycle in an autoclave. During the first heating ramp and the first isothermal, the component is initially under vacuum only. With the start of the second heating ramp, the pressure in the autoclave increases while the vacuum in the vacuum bag decreases. The active principle of this pressure profile is the removal of volatiles and air by the end of the first isothermal. Additional autoclave pressure during these steps (1st heating ramp and 1st isothermal) would have a negative effect, trapping air and other volatiles. After the first isothermal, the consolidation of the component begins with the pressure build-up; it is facilitated by low resin viscosities.

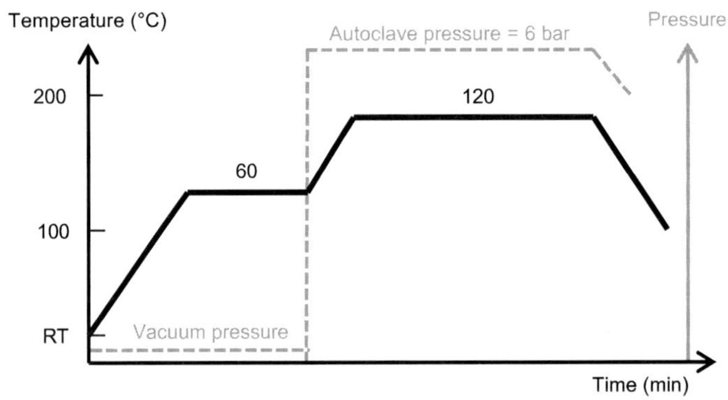

Figure 5.6 Conventional pressure profile during autoclave curing

However, the utilization of the pressure profile described above is not without controversy in practice. Resin consolidation cannot be achieved if the resin begins to gel during the first isothermal. Then the pressure in the autoclave will not be able to reduce porosity or increase component quality. In addition, unpressurized heating of components in the autoclave is rather inefficient, because heat transfer is better with higher pressure in the autoclave (as described above). Another issue of the described pressure profile is related to the hydrostatic resin pressure, which represents the actual pressure in the resin and differs from the pressure in the autoclave. The hydrostatic resin pressure is lower than the autoclave pressure because of the load bearing capacity of the fibers. It determines the solubility of volatiles in the matrix. If the hydrostatic pressure is lower than the vapor pressure of the volatiles, they separate from the matrix and form porosities. If the prepregs are under vacuum only, i.e., there is no autoclave pressure, the hydrostatic resin pressure is very low or even negative. At high temperatures and long time periods, these are ideal conditions for the formation of voids and porosities [2]. In order to avoid these problems, it is common practice to build up a certain pressure in the autoclave (approx. 60 to 85% of max. pressure) at the beginning of the curing cycle. However, maximum pressure is applied only after the first isothermal (Figure 5.7).

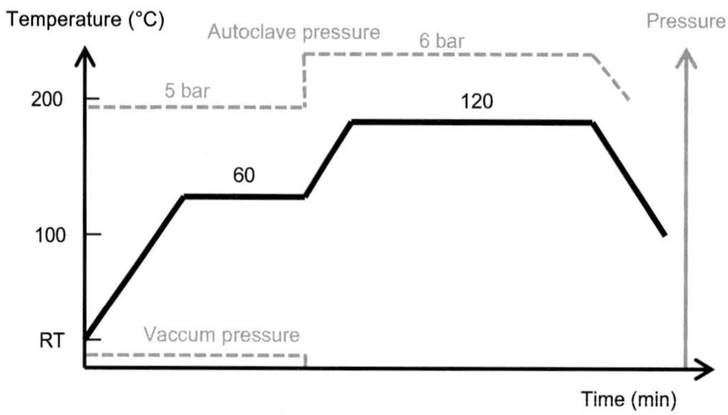

Figure 5.7 Pressure profile with pressure build-up at the beginning of the curing cycle

The timing of pressure build-up has a major effect on the final component properties. Maximum pressure should be applied as soon as the resin viscosity in the topmost layer is low enough to facilitate resin flow. If pressure is applied too early, the resin loss (from the prepregs into the vacuum bag) may become too high. If pressure is applied too late, high resin viscosity may cause insufficient resin flow. The correct timing for pressure application is also dependent on other parameters: it should be earlier the higher the autoclave pressure and the steeper the heating ramp [3].

Compared to all other curing technologies currently available, curing of prepregs in an autoclave provides the best mechanical properties in the resulting components. Air or volatiles in solution or entrapped in the matrix are more effectively transported to the vacuum connection. The gas volume that cannot be removed from the prepreg is reduced proportionally to applied pressure. The process pressure also increases the solubility of volatiles in the resin, which in turn reduces component porosity.

However, the capital investment and installation costs for autoclaves are significantly higher compared to other technologies. The high manual effort and tool and operating costs contribute to the high production costs of components cured in autoclaves.

5.3 Oven Curing

In general, an oven is nothing but a pressure-less autoclave. For curing in an oven, prepregs are again arranged in a vacuum bag and cured at high temperatures. Circulating air ovens are utilized to ensure sufficient gas circulation required for the uniform heating of the components. The maximum component size is determined by the oven size.

The effects of heating ramp and curing temperature on the component quality are the same as those described in Section for autoclave curing systems, and therefore will not be discussed further here. As mentioned earlier, convective heat transfer is less efficient in a pressure-less state than under elevated pressure. Therefore, oven curing requires longer heating times than autoclaves. The temperature distribution in an oven-cured component is also more heterogeneous.

Consolidation of prepregs in an oven occurs under normal pressure (1 bar) and is considerably lower than in an autoclave. The curing of conventional prepregs in an oven typically results in components with reduced mechanical properties. Oven-cured components exhibit porosities of approx. 5 to 10 %. These values are significantly higher than those achievable by autoclave technologies (approx. 1 to 2 %).

However, production costs for oven-cured components are significantly lower than those for autoclave curing because of their lower capital investment, installation, and operational costs. They are the main motivation for countless development efforts in the area of oven curing currently under way. These studies mainly focus on novel resin systems that allow for better fiber wetting and component consolidation under normal pressure. For example, special resin systems based on bismaleimide (BMI) that exhibit excellent mechanical properties and significantly lower porosities are currently tested for oven curing applications [4].

5.4 Quickstep™ Technology

Convection (air or inert gas) is the major mechanism of heat transfer in autoclaves and ovens. The heating ramp is rather restricted by the low thermal conductivity of gases, which results in longer curing cycles. Quickstep™ technology utilizes a liquid rather than a gas as the heat exchanger, as seen schematically Figure 5.8. A flexible membrane separates the component from the liquid.

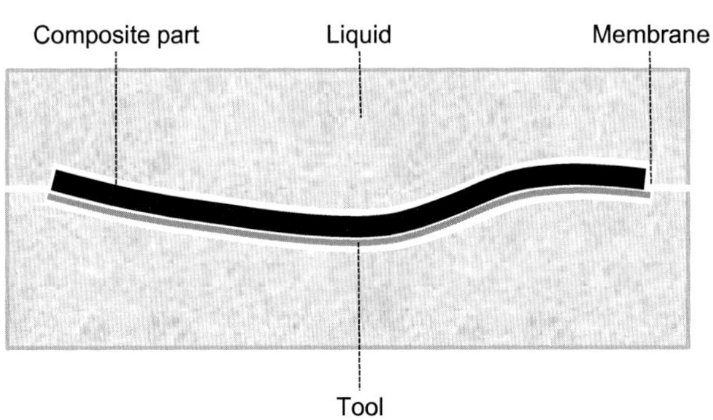

Figure 5.8
Schematic of Quickstep™ technology

Quickstep™ technology allows for steeper heating and cooling ramps (up to 20 °C/min) than achievable utilizing gases (autoclaves or ovens) because the thermal conductivity offered by liquids is higher. Thus, curing cycles become shorter while the isothermal steps remain identical to those required for autoclave and oven processes [5], as seen in Figure 5.9. Cycle time reductions of approx. 43 % for a typical prepreg component were reported for this technology [6].

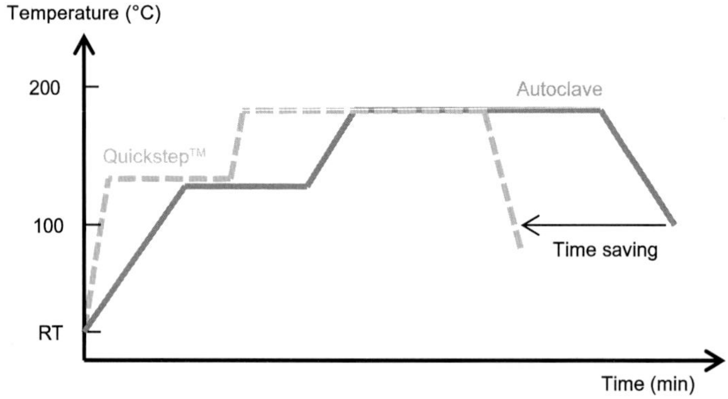

Figure 5.9 Schematic of curing cycles for autoclave and Quickstep™ technology

Another advantage of Quickstep™ technology over autoclave and oven technologies is the fact that the risk of thermal decomposition during the curing cycle is reduced. The heat released from the component by the exothermic cross-linking reaction can be effectively exchanged through the liquid.

With steeper heating ramps the resin viscosity becomes lower, and because Quickstep™ technology facilitates rapid heating, the resin viscosity is correspondingly lower. This results in better fiber wetting and stronger fiber-matrix adhesion.

Although it is possible to generate pressure through the liquid, the achievable differential pressure reaches only approx. 0.1 to 0.3 bar. As a result, the prepreg layers are not as well consolidated, which leads to reduced mechanical properties compared to those achievable with autoclave curing. In addition, the flexible membranes utilized with Quickstep™ technology limit the type of components processible to those with relatively simple geometries.

5.5 Press Molding/Forming Processes

With press molding/forming processes, individual prepreg layers, or a laminate component that has not yet been cured, are placed directly in a tool, after which the press is closed. Here, heat transfer is very efficient, because there is direct contact between the heated tool and the laminate. The tool consists of two parts (female/positive and male/negative part) and is typically made of cast metal. The distance between both parts defines the component thickness. Figure 5.10 shows a press and the respective tool for the manufacture of a prepreg shell.

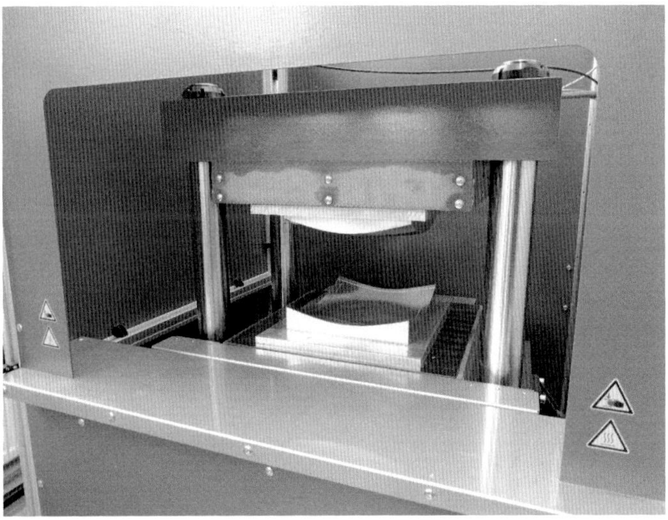

Figure 5.10 Press molding for the manufacture of fiber reinforced components [Courtesy: SKZ – Das Kunststoff-Zentrum]

Hydraulic pressure is used to consolidate the laminate. The temperature required to initiate curing of the component may be introduced by different means, including electric cartridge heaters or water vapor. The prepregs are in direct contact with the temperature-controlled tool surface, therefore heat transfer is faster here than in processes that are based on convection, such as autoclave and oven.

Typically, isothermal cycles are preferred for the curing of prepregs, because the press tools exhibit high thermal inertia and make it difficult (and costly) to implement fast heating and cooling ramps.

Press/forming processes allow very narrow component tolerances. They also facilitate a high degree of automation. The curing cycles can be precisely traced and controlled, which allows for the manufacture of components with high quality and reproducibility. Very high tool pressures (30 to 60 bar) can be achieved, which results in excellent component consolidation and low porosity. Therefore, the mechanical properties exhibited by components manufactured by press molding/forming processes are very good.

Because of their high capital investment and tool costs, press molding processes are particularly suited for the manufacture of high volume, high production rate parts. For smaller production runs, tools made from fiber reinforced plastics or electroformed nickel can be utilized.

The disadvantages of this technology include the limitations of component geometry and the lack in flexibility. Even small changes in component design require the fabrication of new tools, which results in high capital investment costs.

5.6 Induction Processes

Induction curing technology is a derivation of press molding processes and is marketed by RocTool, SA, under the trade names Cage System® and 3iTech®. Induction processes also utilize a press, but in terms of tool technology there are major differences compared to conventional press forming processes. Induction will heat only the surface of the cast tool, and the cooling water channels are also placed close to the tool surface, which facilitates fast cooling of the component.

Heating and cooling only the surface of the cast tool makes induction processes rather energy-efficient. The tool surface is heated by electromagnetic induction [7], which allows for heating ramps of up to 5 to 10 °C/s (300 to 600 °C/min). An induction coil placed around the tool generates an electric field and induces an electric loop in each tool half, as seen in Figure 5.11. The current flow through electrical conductors creates heat (Joule effect).

The current flows through the entire surface of the tool; however, heating is desired only in those parts of the tool that are in direct contact with the laminate to be cured. Therefore, the outside tool surfaces are coated with a metal that exhibits a low Joule effect. Here, copper is typically used because of its high electric conductivity and non-magnetic behavior.

On the tool surface, a strong Joule effect is required for fast heating; therefore, a material with high electric resistivity is used here. In addition, the material needs to exhibit relatively high magnetic permeability, because that results in high relative tool performance, i.e., the material heats in relatively low magnetic fields. Preferred tool surface materials are steel alloys with permeability numbers of Mu_r= 50 or higher [8].

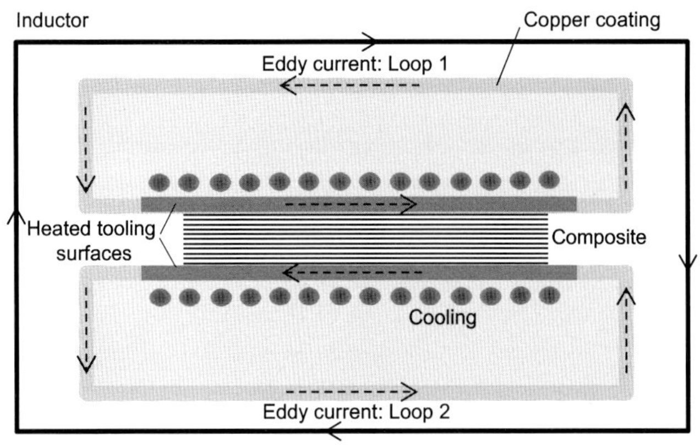

Figure 5.11 Schematic set-up of an induction process tool

Like press forming processes, induction curing technologies are utilized for medium to large production runs. Even small changes in tool design are almost impossible because of the positioning of the cooling channels. Also, only components with very simple geometries (even simpler than for press forming processes) can be manufactured by induction curing. If the components had more complex geometries, the temperature distribution at the tool surface would be inhomogeneous, which in turn would result in inhomogeneous curing of the prepreg layers.

The main advantages of this technology, compared to press technologies, are its energy efficiency (only the tool surface being heated) and the option to very rapidly heat up the tool surface. With the induction tool mounted in a press, it is possible to manufacture components with high degrees of consolidation and low porosity.

5.7 Microwave Curing

Microwaves are high-frequency electromagnetic waves which follow Maxwell's equation similar to light. Microwaves are defined as the frequency range between approx. 10^8 and 10^{12} Hz in the electromagnetic spectrum and microwave curing technology has established a standard frequency of 2.45 GHz.

Different materials absorb and reflect microwaves at different levels. The absorption of microwaves, and thus the heating of a material, is mainly based on two mechanisms. On the one hand, free moving charges are able to absorb microwaves by induction. The formation of eddy currents creates heat that propagates through the irradiated material. This mechanism is observed in metals, semiconductors, and solutions with free ions. On the other hand, polar molecules, which are intrinsically neutral, react to microwave irradiation by constantly oscillating with the alternating field, thus causing intramolecular friction. Materials exhibiting this behavior include water and a variety of plastics. The effect of creating heat through intramolecular friction in an electric field is also called dielectric loss. The higher the dielectric loss, the more pronounced a material's reaction to exposure to microwaves and the shallower the depth of penetration.

A key challenge for microwave process technology is the generation of uniform, homogeneous electromagnetic fields, yet at the same time this is a technical prerequisite for many industrial applications. In a conventional kitchen microwave oven, the content to be heated is rotated to compensate for the inhomogeneity of the electromagnetic field. However, in many technical applications this is not possible, and sometimes it is even undesirable. Industrial microwave ovens utilize several magnetrons that turn on and off successively in order to generate a homogeneous electromagnetic field. In other words, here the electromagnetic field rotates rather than the component. This technology is marketed by Vötsch Industrietechnik GmbH, Germany, under the trade name Hephaistos Microwave System (Figure 5.12). Standing wave structures are significantly reduced by a hexagonal applicator with a total of 24 magnetrons. The exceptional homogeneity of the field is achieved primarily by the oven geometry.

Figure 5.12 Hephaistos microwave system for curing of fiber reinforced components

The dielectric loss of resin systems is just sufficient to ensure good response to microwaves. With continuing curing reaction, microwave reactivity will decrease. This means that at constant microwave output, heating of the material will decline over time (as the reaction proceeds). This effect provides "self-protection" from overheating for the resin and allows for homogeneous curing reactions. Figure 5.13 shows the dielectric loss factors for a variety of materials.

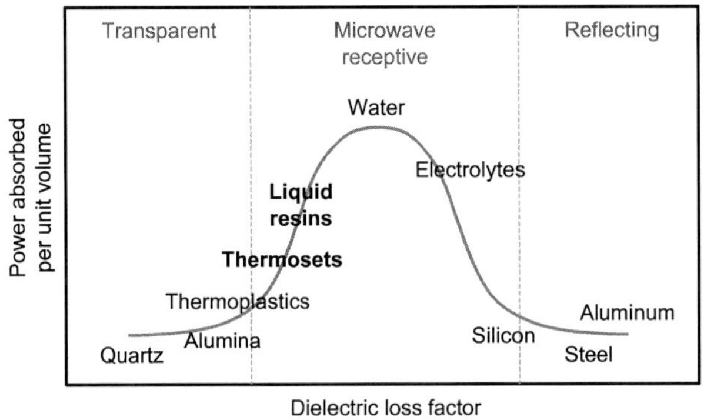

Figure 5.13 Resins exhibit higher dielectric loss factors than cured thermosets [9]

Compared to autoclave and oven curing technologies, the major advantage provided by microwave curing is the fact that it heats only the component while the surroundings remain at room temperature. This makes it an energy efficient process with energy costs significantly lower than those for autoclave curing. The process also allows for heating ramps of up to 100 °C/min. However, active cooling of the component is not possible.

A general problem commonly observed in oven-cured laminates is also seen when utilizing microwave curing. Because the components are not under pressure during curing, they exhibit high porosity, and the mechanical properties of microwave-cured components are therefore inferior to those of autoclave-cured components [10].

Corners and edges experience higher heating rates because the energy introduced in the material depends on the irradiated area. Thermocouples may also lead to temperature variations, a fact to be considered in particular when very homogeneous heat distributions are required. Commercially available prepregs can be cured using microwaves; however, for high-performance components that require very homogeneous curing profiles the process still needs to be further developed. Currently, microwave curing has not yet established itself.

5.8 Electron Beam Curing

In the field of manufacturing technologies, a non-thermal, out-of-autoclave process with significantly reduced curing times promises the highest potential for cost reduction. On a technical level, electron beam curing of fiber reinforced composites offers the best chance to cure fiber reinforced components within minutes at room temperature.

Photochemistry (absorption of light and UV radiation) is based on the interaction between a photon and an atom or molecule. The interaction excites a valence electron and lifts it to a higher energy level. High-energy electron beam radiation may even completely remove electrons from an atom or molecule and thus create ions or radicals that facilitate chemical reactions/cross-linking. While UV curing of epoxy resins for coatings is a technically established process, it is limited to thin components with low levels of filler content. Fiber reinforced components cannot be manufactured by UV curing. However, electron beam radiation, with its higher level of energy, is a suitable process to initiate and promote curing reactions in fiber reinforced components. Components with thicknesses up to 30 mm can be cured utilizing this process.

State-of-the-art in the manufacture of fiber reinforced materials are thermally curing, amine cross-linking systems. In contrast to the step mechanism of thermal reactions, the photochemically induced cross-linking reaction of epoxy resins is a chain reaction (initiated by a photoinitiator). Epoxy resins cross-link via either a cationic or via a combined cationic/radical chain reaction mechanism.

Cationic polymerization is a "chain reaction" and therefore proceeds at high rates than "step reactions". A component will cure in only minutes, compared to hours required for conventional, thermally reactive systems. The growing macro-cation undergoes chain-extending reactions with the monomers. The rate of this reaction is determined by the interaction between the ionic chain end and the respective counter-ion. The rapid reaction rates of electron beam induced cross-linking reactions are the result of fast gel point attainment. The lack of thermal energy and the formation of macromolecules cause a rapid increase in viscosity and thus an early diffusion restraint. Other than for thermal curing processes, the literature does not yet provide data with respect to viscosity development during electron beam curing.

The cross-linking reaction takes place in the course of three reaction steps [11] (Figure 5.14):

a) *Formation of Brønsted acid:* In order to initiate polymerization it is necessary to form a Brønsted acid through energy impact (via photoinitiators). The formation of the Brønsted acid is the characteristic part of electron beam curing, because only this sub-reaction requires electron irradiation. All further reactions theoretically proceed independent of irradiation.

b) *Initiation of ring-opening polymerization:* Cationic polymerization is initiated by the reaction between electrophilic initiators with monomers that act as electron donors. In the case of electron beam curing of epoxy resins, the Brønsted acid acts as the electrophilic initiator. Here, it is therefore the actual initiator for cationic cross-linking. The active hydrogen ions of the Brønsted acid cause protonation of the epoxy ring and thus initiate ring-opening polymerization.

c) *Chain growth and termination:* In the simplest case, cationic polymerization reactions propagate by the repeated addition of macromolecules to carbocations or onium ions under preservation of the ionic character. Cationic polymerization reactions are characterized by numerous termination and transfer reactions. One of the most important termination reactions is the result of the interactions between the growing polymer chain and water or alcohols. Both contaminations act as chain terminators because of their hydroxyl groups (R-OH).

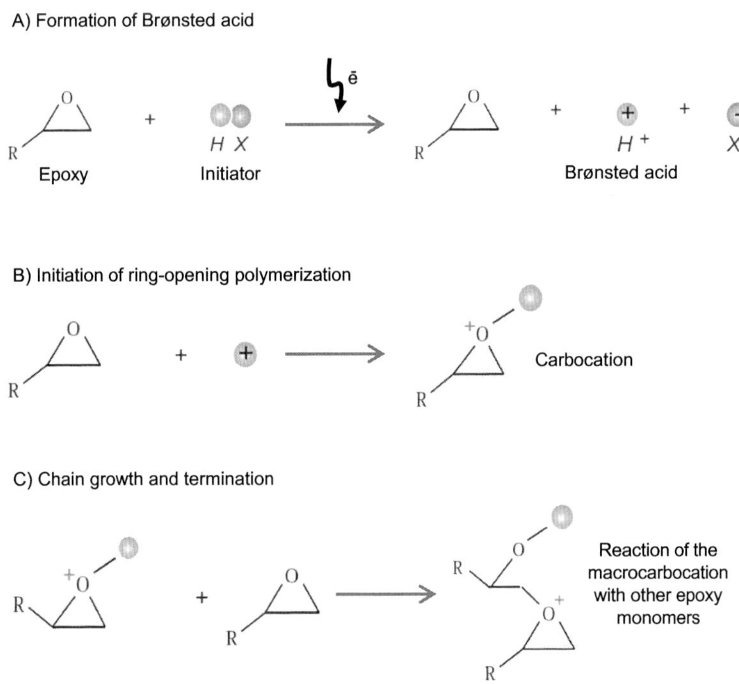

Figure 5.14 Schematic process of electron beam curing of epoxy resins

Non-thermal, out-of-autoclave electron beam curing offers the following advantages over autoclave technology [12]:

- Reduction in process costs because of shorter cross-linking time and elimination of autoclave technology. The literature reports an overall cost savings potential of 10 to 40%.
- Electron beam curing allows for the manufacture of larger components because the component size is no longer limited by the size of the autoclave, but only by the size of the transport carriage used to transport the component through the electron beam line. Increasing the size of this carriage is more cost effective than up-scaling of an autoclave.
- Curing does not require the otherwise necessary high temperatures so that components can be cured at room temperature. The component therefore experiences lower thermal loads, which in turn reduces the formation of thermally induced residual stress in the component. Curing at room temperature also facilitates the use of cheaper tool materials, such as wood or plastic.

Irradiation and curing of components by electron beam technologies takes place in specifically designed facilities that exhibit radiation efficiencies of up to 10 MeV. Figure 5.15 schematically describes an electron beam line.

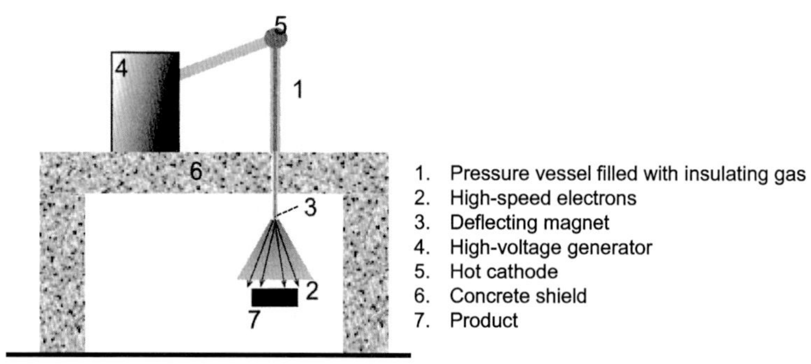

Figure 5.15 Schematic of an electron beam line for curing of fiber reinforced components

In a study, the energy efficiency of a 50 kW electron beam line with a curing dose of 100 kGy and 70% electron beam exposure was compared to a conventional autoclave [13]. The results are shown in Table 5.1. The energy consumption of a typical accelerator utilized for the curing of fiber reinforced materials is only approx. 10% of the energy required for curing by autoclave (cycle time: 4 h).

Table 5.1 Energy Efficiency of Electron Accelerator Compared to Traditional Autoclaves [13]

Equipment type	Equipment specification	Total energy (kW/h)	Capacity (kg/h)	Energy efficiency (kWh/kg)
10 MeV Electron beam	50 kW	400	1800	0.22
Autoclave	12.2 m length 2.4 m diameter	480	273	1.76
Autoclave	15.2 m length 7.6 m diameter	7660	2730	2.81

Conventional, thermally curing prepregs based on epoxy-amine systems cannot be cured by electron beam radiation. This process utilizes specifically developed resin systems based on epoxy resins and photoinitiators (approx. 1 to 3 wt.%). Electron beam curing prepregs were first tested in the 1990s for military applications [14].

In 2001, Air Canada utilized electron beam curing of prepregs for the repair of fairings of the Airbus A320 cargo plane. Here, prepreg plies were laid up and compacted on the component to be repaired. A vacuum bag was used to remove the air between the plies. Subsequently, the stack was cured by electron beam irradiation.

Although electron beam curing provides a number of advantages over autoclave curing, the technology is not yet ready for the market. From a material perspective, one reason is the fact that their high cross-linking densities result in electron beam cured thermosets exhibiting very low fracture toughness. Also, the degree of fiber wetting is insufficient because of the fast curing reaction, which in turn makes it difficult to achieve strong fiber/matrix adhesion. The mechanical properties of electron beam cured components do not match those of autoclave cured components. The relatively complex process requires profound knowledge regarding the effect of a variety of parameters (such as dose, energy, number of passes, and irradiation time and rate) on the curing process, which so far has been investigated in only few studies [15]. The current state of electron beam curing technology requires intensive research and technological development.

5.9 References

[1] D. Purslaw, R. Childs. *Composites*. 1986, 17, p. 757.

[2] F.C. Campbell. *Manufacturing Processes for Advanced Composites*. Oxford: Elsevier, 2004.

[3] P.K. Mallick. *Fiber-Reinforced Composites: Materials, Manufacturing and Design*. Boca Raton: CRC Press, 2008.

[4] C. Ridgard, J. Russell. *2011 High-Performance Resins/Highlights*. High Performance Composites. 2011, November, pp. 30–32.

[5] V. Coenen, M. Hatrick, H. Law, D. Brosius, A. Nesbitt, D. Bond. *A Feasibility Study of Quickstep Processing of an Aerospace Composite Material*. Paris: SAMPE Europe International Conference, 2005.

[6] M. Kaiser, C. Garschke, B. Fox, C. Weimer, K. Drechsler. *Out of Autoclave Manufacture of Structural Aerospace Composite Materials*. Paris: SAMPE Europe International Conference, 2007.

[7] D. Vink. *Induktionsheizung: Highspeed-Werkzeugtemperierung für Qualitätsteile*. (Online) (accessed 26.11.2012) http://www.industrieanzeiger.de/erp/-/article/32571342/37715087/Mit-Induktion-zu-Hochglanz/art_co_INSTANCE_0000/maximized/ (accessed 6/14/2014).

[8] RocTool. *Inductive Heating Modelling with a Temperature-Controlled Material*. Paris: COMSOL Multiphysics User's Conference, 2005.

[9] E.T. Thostenson, T.W. Choud, *Composites Part A*. 1999, 30, pp. 1055–1071.

[10] C. Nightingale, R.J. Day. *Flexural and Interlaminar Shear Strength Properties of Carbon Fibre/Epoxy Composites Cured Thermally and with Microwave Radiation*. Composites Part A. 2002, 33, pp. 1021–1030.

[11] F. Wolff-Fabris, V. Altstädt, U. Arnold, M. Döring. *Electron Beam Curing of Composites*. Munich: Carl Hanser Verlag, 2011.

[12] A.J. Berejka. Radiation Physics and Chemistry. 2002, 63, p. 551.

[13] V.S. Ivanov. *Radiation Chemistry of Polymers*. Utrecht: VSP BV, 1992.

[14] J.D. Farmer, C.J. Janke. *The Electron Beam Cure of Fibreglass/Epoxy Prepregs – Final Report*. Oak Ridge: USAF Advanced Composite Program Office, Oak Ridge National Laboratory and Atomic Energy of Canada Ltd.

[15] F. Wolff-Fabris. *Toughness Modification of Electron Beam Curable Epoxy Matrices for Carbon Fibre Composites*. Bayreuth: Department of Polymeric Materials, University Bayreuth, 2009.

6 Tooling Technology

Hauke Lengsfeld and Hilmar Apmann

This chapter will describe the tooling used for and the concepts underlying the forming and curing of prepregs. Exemplary concepts will be explained in detail to illustrate the differences in requirements.

Prepregs are two-dimensional, semi-finished products delivered to the manufacturer in the form of rolls. It requires dimensionally stable tools to transfer these pliable materials into a predetermined contour and to cure them. In principle, the fiber reinforced material is generated in the moment when matrix (resin) and fibers are brought in contact, i.e., the manufacturer creates the material while processing it. The forming operation occurs virtually in situ with the creation of the material, which represents the fundamental difference compared to conventional materials, such as wood or metals. Although these materials can also be formed by appropriate tools, they are not created during this process. The accuracy and reproducibility of the forming operation substantially affect the performance of fiber reinforced materials because even minor disruptions in fiber orientation or porosities lead to reduced strength and stiffness of the material.

The term "tooling" is often used with respect to the manufacture of fiber reinforced components and describes tools that form the component while it cures.

Figure 6.1, left, shows exemplary tooling used for the manufacture of fiber reinforced components from prepregs among other materials. The forming operation begins with the cutting of the prepregs and continues through hot forming and subsequent curing (Section 4.6). In some applications, specific tools are used to correctly position the prepreg plies. Tools are used across the different industries that process and utilize composite and prepreg materials (Figure 6.1, right).

Tooling selection depends on the processing technology, number of units, and the size of the components to be produced. Also to be considered are requirements with regard to surface quality and dimensional accuracy. Tooling applications for the manufacture of composite components include the aerospace, wind energy, boat building, and automotive industries (Figure 6.1).

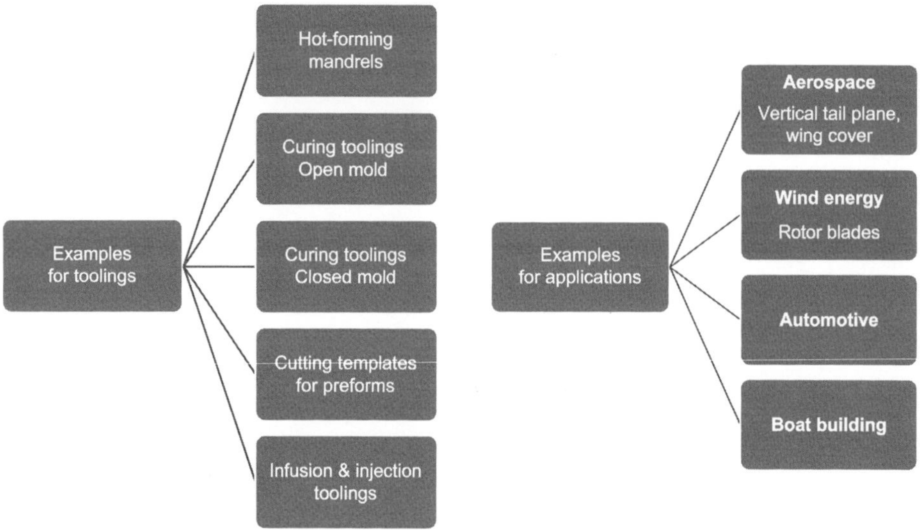

Figure 6.1 Examples of toolings (left) and applications (right)

■ 6.1 Requirements

Tools for the processing of prepreg materials have to meet a number of requirements and characteristics that are essential to ensure the manufacture of high-quality composite components.

Table 6.1 lists the essential requirements and characteristics.

Table 6.1 Requirements for and Characteristics of Prepreg Tooling

Requirements	Characteristics of Tooling
▪ Thermal characteristics: convective heat transfer into the laminate should be fast and efficient	▪ Preferably low heat capacity for short curing cycles ▪ Thin-walled tools
▪ Contour accuracy under process conditions	▪ Material with low thermal expansion or tooling with precalculated geometry change due to thermal expansion
▪ High surface quality of tooling and composite part	▪ Wear resistant tooling surfaces for durability ▪ Good demolding characteristics
▪ Good laminate quality (void-free)	▪ Permanently vacuum tight
▪ Good handling characteristics	▪ Easy cleaning ▪ Repairability ▪ Low weight
▪ Cost effectiveness	▪ Low tooling manufacturing cost ▪ Good availability of tooling material ▪ High durability under cyclic temperature and pressure loading

The thermal characteristics of a tooling utilized for prepreg curing affect not only the quality of the laminate, but also the time required for curing, e.g., in an autoclave. The heavier the tooling, the higher the inertial thermal mass during the heat-up period. This leads to slower heating of both tooling and component. The tooling material also determines the heat transfer into the laminate. It is therefore appropriate to use a light and/or thin-walled tooling in order to keep the curing cycle short and more cost-effective.

The contour accuracy of a tooling during the curing process is the most crucial characteristic to ensure the manufacture of high-quality composite components. It implicates that the tooling dimensions are allowed to change only within defined limits during the curing and cooling phases. The dimensional changes depend on the used material and its coefficient of thermal expansion, CTE. Therefore, the thermal expansion behavior of tooling and component have to be closely aligned. Otherwise, the component (during a state of incomplete curing) may adapt to the changed (incorrect) contour of the tooling and then complete the curing cycle. While the effect of tooling contour changes is minimal in flat laminates, its importance increases significantly in multiple-curvature composite structures. For example, in the worst case the composite component may shrink onto the tooling (positive tooling) and/or the shrinking may cause the tooling to warp or buckle (negative tooling) (Figure 6.2), or it may be rendered defective because of crack formation in the laminate (see also Chapter 8).

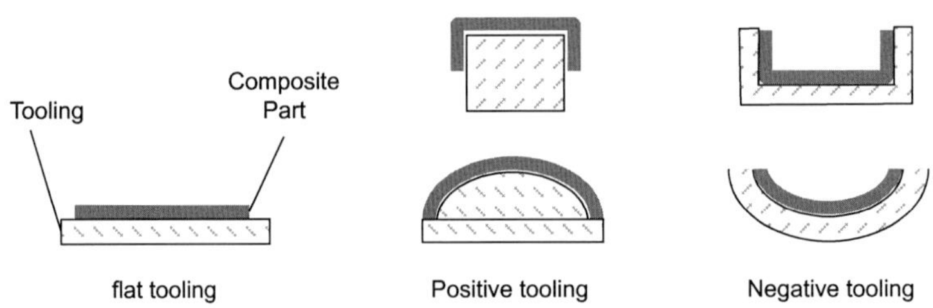

Figure 6.2 Tooling contours

In order to achieve non-porous components and thus good laminate quality, it is important to make sure that the tooling is vacuum-sealed. Because curing of prepregs in an autoclave occurs under high pressure (7 to 10 bar; for generation of pressure differential, see Chapter 5), it is important to make sure that the autoclave pressure does not penetrate into the vacuum bag. The prevention of internal voids in a laminate is typically accompanied by high surface quality on the tooling side of the final component. The latter minimizes time-consuming surface finishing efforts such as sanding and priming. As a consequence, the quality of the tooling also affects its cost effectiveness for the manufacture of composite components.

Additional important requirements for a tooling are favorable mechanical handling characteristics, such as good demolding and cleaning behavior, the possibility to make surface repairs, and low net weight; some of these are interrelated. The simpler it is to demold a component and to clean the tooling surface, the less wear the tooling is subjected to, and thus the less likely is the need for repairs. Low tooling net weight facilitates not only fast curing, but it also reduces handling costs, such as moving it by crane, rail system, or on a simple roll car. Implementation of a crane system is typically more cost intensive than utilization of a rail system.

Additional economic aspects are tooling manufacturing costs and service life, which depend on the tooling material, its availability, and its durability. The durability of the selected tooling material is of particular importance for autoclave toolings that have to withstand high cyclical temperature and pressure loads in order to avoid premature high wear.

6.2 Tooling Materials

A variety of different materials is available for the manufacture of tools for the forming and curing operations of prepregs materials; these materials can also be combined, depending on the area of application.

Table 6.2 provides an overview of common tooling materials that will be described in more detail in the following.

Table 6.2 Examples of Tooling Materials

Material	Description	Type (examples)
Metals	▪ Steel ▪ Aluminum ▪ Nickel-steel alloy (Ni-36)	▪ S235 JR, S355 JR ▪ ALMG 3/3.3535 ▪ INVAR 36, Pernifer 1.3912
CFRP (epoxy resin)	▪ Fabric prepreg ▪ Quasi-isotropic prepreg mat	▪ Cycom® 7620, ▪ HexTOOL® M81
CFRP (BMI or BOX resin)	▪ Fabric prepreg ▪ Quasi-isotropic prepreg mat	▪ Duratool® 5270 ▪ HexTOOL® M61 ▪ Toolmaster BetaPreg
CFRP foam	▪ Carbon foam	▪ Touchstone CFoam® 20
GFRP	▪ Dry fabric + resin, processed by resin infusion or hand lay-up	
Other materials	▪ Wood ▪ Epoxy tooling boards ▪ Cellular concrete	▪ OBO-Plywood, RETIstab ▪ Necuron, Rampf WB700, ▪ OBO-Modulan, TB650 Series ▪ Ytong

6.2.1 Metals

Metals are the easiest and most commonly used class of materials for the construction of tools for prepreg processing. This is due to their easy availability as well as their high load capacity. In addition, metal curing tools stand out for their robustness and thus high structural durability so that they can be utilized without difficulty for more than 1000 cycles. The surface of these tools is resistant to organic solvents and release agents, and even damage, such as scratches and dents, can typically be repaired easily.

Steel, aluminum, and ferronickel alloys are generally utilized, with steel and aluminum being the materials of choice because of their price and durability. Both materials exhibit a high coefficient of thermal expansion (Table 6.4) that needs to be taken into consideration during design of the tooling, in particular for high curing temperatures, in order to ensure its dimensional accuracy. Depending on tooling and application, thermal expansion may be desired, e.g., in order to facilitate for the finished component to shrink off the tooling during cooling (Chapter 8). Contour accuracy of 0.3 to 0.4 mm, even with large toolings (e.g., 20 × 5 m), can be achieved with metal tools at room temperature.

Figure 6.3 Steel tooling for the manufacture of rotor blades [Courtesy: Premium Aerotec GmbH]

A disadvantage of steel is its high weight, which complicates handling and impedes heat transfer. Metal curing tools are generally manufactured by contour milling, but also by preforming of sheet metal. In order to ensure the created geometry, in particular for a large tooling (e.g., thick profiled sheet), even at high temperatures and/or during handling (e.g., by crane), the geometry is often stabilized by a stiff sub-structure ("egg carton" structure). This sub-structure also facilitates safe handling and installation, e.g., of vacuum lines and hoisting points to lift the tooling via crane.

Figure 6.4 Egg carton sub-structure for tooling stabilization
[Courtesy: Premium Aerotec GmbH]

The ferronickel alloys used are special nickel steels with 36 % nickel and 64 % iron content (types such as Invar 36, Permalloy, and Ni36). This material exhibits a particularly low CTE that is almost identical to those of CFRP components, even at high curing temperatures (180 °C). A disadvantage, similar to steel, is the high net weight of these high-density materials. In addition, machining of tooling from ferronickel requires special knowledge of these materials because of the toxicity of the released dust and vapors (during welding). Surface damage, such as scratches and dents, occur relatively easily because of the high material ductility and their repair requires high effort. Like aluminum and steel, Ni36 is resistant to organic release agents and solvents. However, when exposed to water-based release agents it tends to corrode. Ni36 is utilized in particular in the aerospace industry for curing tools for large, highly contoured composite structures, such as wings (35 × 8 m) or fuselage segments (17 × 4 m).

Figure 6.5 Invar tooling for the manufacture of CFRP wing fairings, dimensions 20.5 × 4.8 m
[Courtesy: Premium Aerotec GmbH]

Nickel-galvano tools represent a special form of Ni36 tools; they facilitate the molding of composite components with very smooth, Class-A surfaces.

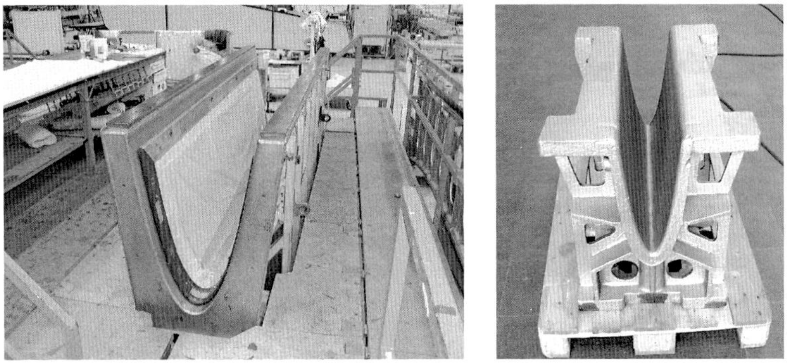

Figure 6.6 Nickel-galvano tooling (left) und cast nickel tooling (right) [Courtesy: Premium Aerotec GmbH]

6.2.2 Fiber Composite Materials GFRP/CFRP

Compared to metals, fiber reinforced materials such as GFRP and CFRP provide a variety of advantages for tool construction. The much lower density of fiber reinforced plastics (FRP), in particular of CFRP, allows for the construction of considerably lighter tools. The lower net weight of the tooling makes handling easier and generally facilitates more uniform, faster heating of tooling and component to be cured.

Figure 6.7 CFRP tooling made from fabric prepreg [Courtesy: Lufthansa Technik GmbH]

Because of the slightly negative coefficient of thermal expansion of the carbon fibers that significantly reduces the thermal expansion of the matrix, CFRP tools exhibit much lower CTEs than comparable metal tools. Often, their CTE is almost identical to that of the component so that there is virtually no thermal expansion between tooling and component during the curing cycle. The expansion of both tooling and component is heavily dependent on their respective laminate structures. Figure 6.8 shows the coefficients of thermal expansion for CFRP and various metals. The curves clearly indicate how the CTE for CFRP materials changes as a function of laminate structure. Fiber reinforced plastic tools are typically designed with a quasi-isotropic structure (mainly fabric structures). However, the structure should match the laminate structure and geometry of the prepreg component (see also Chapter 8) in order to avoid thermal strain between both.

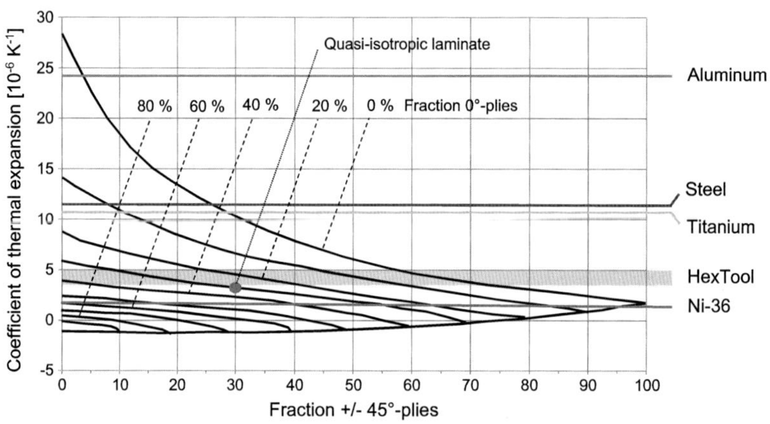

Figure 6.8 Coefficient of thermal expansion of metals and CFRPs as a function of laminate structure

Significant differences in CTE, such as between steel/aluminum and CFRP components, may be advantageous for the demolding of positive toolings (Figure 6.2). Here, the tooling exhibits a higher CTR than the CFRP component, and therefore the tooling will contract to a higher degree than the CFRP component during cooling, thus shrinking free. Now, final component and tooling can be easily separated.

Compared to their metal counterparts, tools made from FRP exhibit considerably shorter service lives, depending on service temperature and period of application as well as resin matrix used. While metal tools easily reach 1000 cycles and more, FRP tools often can be utilized for only several hundred cycles. The FRP matrix will degrade under the high temperatures required for prepreg curing in an autoclave and the increased partial oxygen pressure. Epoxy resins in particular exhibit strong oxidative decomposition of the matrix under high temperatures (150 to 200 °C). This leads to the formation of micro-cracks in the surface (Figure 6.9) and to leakages (creeping process) so that the tooling can no longer hold sufficient vacuum. Components manufactured with such tools often show voids in the laminate. Other typical signs of wear include surface abrasion, caused mainly by scratches and/or cleaning operations such as sanding and polishing.

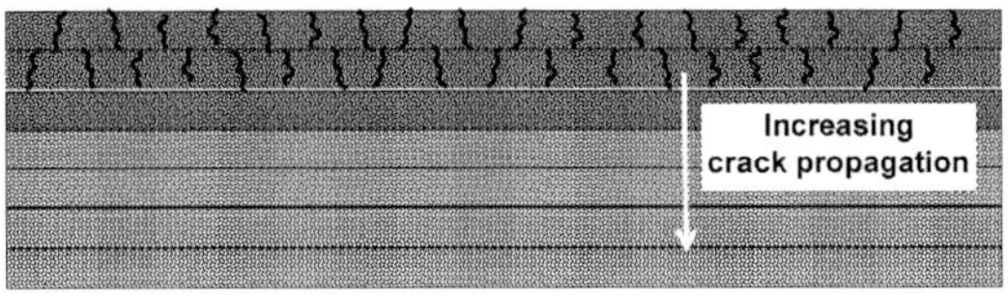

Figure 6.9 Surface micro-cracking of tooling made from epoxy resin

Increased service lives of up to approx. 1000 lay-ups can be achieved with BMI matrix resins. They are cured at temperatures of approx. 240 °C, which is considerably higher than the curing temperature of common epoxy resin-based prepreg systems (120 to 180 °C for complete curing). Because of their high curing temperatures, BMI systems also exhibit high glass transition temperatures (T_g), which lead to good dimensional stability of the tooling. These systems are also resistant to oxo-degradation owing to their chemical composition. Figure 6.10 shows the percentage weight loss of a BMI matrix in comparison to an epoxy matrix over a long-term aging process. The figure shows that the BMI matrix degrades at a considerably slower rate.

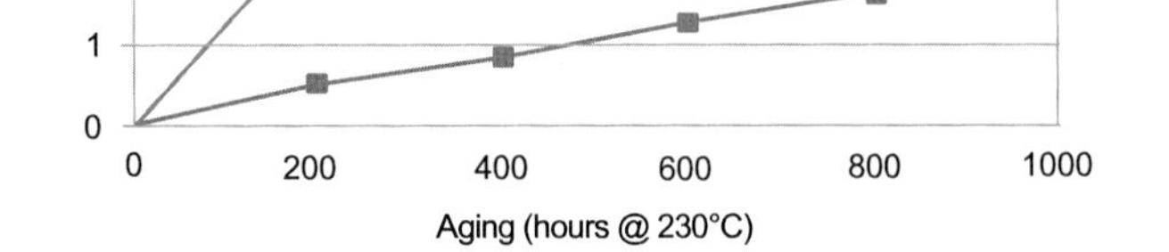

Figure 6.10 Aging of BMI matrix compared to epoxy matrix

Like metal tools, the surfaces of FRP tools are typically not susceptive to release agents, and small damages are easily repaired.

Table 6.3 compares the characteristics of Ni36 with those of CFRP materials with BMI matrix as they relate to their use as tooling materials.

Table 6.3 Comparison of Metals and CFRP as Tooling Materials

Requirement	Objective	Nickel36 alloy	CFRP (BMI matrix)
Cost of acquisition	As cheap as possible	Expensive duplicates	Duplicates: same price as first tooling, material often cheaper than Ni-36 alloy
Surface finish	Direct use of release agents possible	Difficult to mill/trim, established technique	Difficult to mill/trim, established technique
Surface durability	Unlimited	Unlimited given approp- riate maintenance	Unlimited given appropriate maintenance
Durability, number of cycles	1000+ cycles	Unlimited	1000+ cycles
Flexibility for design changes	Material easy to process and to mill/ trim in contour	Difficult to weld and to mill in contour w/o additional annealing	Additional material can be laminated on surface, easy re-milling of surface contour
Tooling modifica- tions	Tooling geometry easy to adapt	Surface tolerances very difficult to maintain when completely re-milling	Easy to modify w/o loss in tolerances
Repairability	Tooling can be repaired with simple and established techniques	Welding and contour re-milling of surface	Lay-up of additional material on the surface and re-milling of contour

Figure 6.11 shows a GFRP tooling used for the manufacture of rotor blade components. Because resin systems with low curing temperatures (up to approx. 80 °C) are often used for these components, the thermal expansion of the tooling is typically low, making GFRP/epoxy systems suitable materials for these applications.

Figure 6.11
GFRP tooling [Courtesy: Grunewald GmbH]

6.2.3 Other Materials

Materials such as wood, cellular concrete, foams, or epoxy tooling blocks are typically not used directly as curing tools for prepregs because they offer neither the required durability nor the temperature stability (Table 6.4). However, wood and epoxy tooling blocks are perfectly suitable as preform and/or hot forming tools for smaller size components at service temperatures up to approx. 120 °C. Another application for epoxy tooling blocks, cellular concrete, and foams is the manufacture of master molds that are used to lay up prepregs for the final production tool. Porous surfaces, such as those of cellular concrete, are sealed with silicone sealing agents prior to being used as a forming tool.

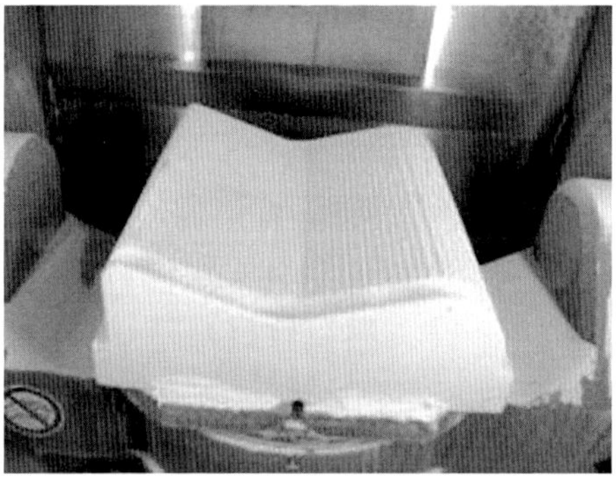

Figure 6.12
Pre-tooling made from milled epoxy tooling block [Courtesy: Grunewald GmbH]

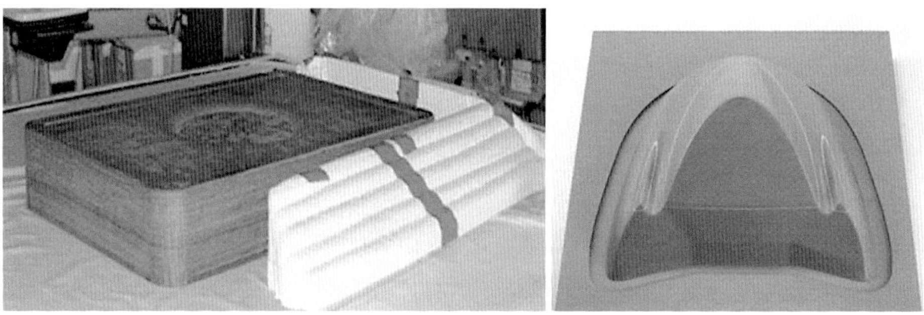

Figure 6.13 Tooling for hot forming made from wood (left), Necuron tooling material (right) [Courtesy: Necumer GmbH]

Table 6.4 provides an overview of selected characteristics of various tooling materials.

Table 6.4 Characteristics of Various Tooling Materials

Material	Type (example)	Density in g/cm³	CTE in 10^{-6}/K	Service temperature in °C	Dimensional accuracy	Durability	Production technology
Steel	S235 JR	7.8	12	up to 200	–	++	Contour milling
Aluminum	ALMG3/3.3535	2.75	24	up to 200	– –	++	Contour milling
NiFe36	Pernifer 1.3912	8.1	1.5	up to 200	++	++	Contour milling of surface
CFRP-epoxy	HexTOOL® M81	1.4–1.6	4-5	up to 180	+	0	Prepreg curing in autoclave + contour milling of surface
CFRP-BMI	HexTOOL® M61	1.55	4.2	up to 200	++	++	Prepreg curing in autoclave + contour milling of surface
CFRP foam	CFoam 20	0.32	5	up to 500	+	++	Contour milling
Epoxy tooling board	Rampf WB700	0.7	40	up to 135	++	–	Contour milling
Wood	OBO plywood RETIstab	0.41–0.48	3 to 5	RT up to 150	0	–	Contour milling
Cellular concrete	Ytong	0.35–0.65	8	up to 250	0	–	Contour milling

– – very bad; – bad; 0 acceptable; + good; ++ very good

6.3 Manufacture of Tools

The techniques used for the manufacture of tools may vary depending on the material utilized and the technology for which the tools are intended. Figure 6.14 shows the fundamental principles for the manufacture of a curing tooling for prepregs starting from a CAD data set. Simple molds, e.g., for hand lay-up in boat building or model making, are made based on two-dimensional drawings rather than CAD data sets.

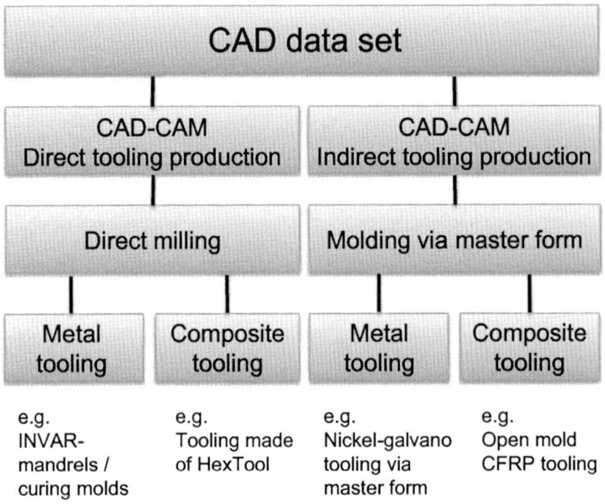

Figure 6.14
Manufacturing principles for forming tools

6.3.1 Direct Manufacturing of Forming Tools

Direct tool manufacturing describes manufacturing methods that directly mill a tooling from the tooling material (metal or CFRP by means of a specifically created CAD model. For this purpose, CAD data are converted into a format that can be read by the milling machine (CAD-CAM chain for 3-D data). The tooling is first welded from individually preformed sheets, stress-relieved by annealing, and sand blasted. Subsequently, the mold surface is machined and honed (surface finish) to create the final contour. With suitable milling machines, this manufacturing method allows for very tight tolerances (contour accuracy of 0.3 to 0.4 mm) of the final tooling. Finally, attachments and accessories, such as the vacuum system, thermocouples, and hoisting points are mounted. Before the tooling is delivered to the customer, it is tested for (leak) tightness and contour accuracy.

Figure 6.15
Tooling made of INVAR for the manufacture of CFRP composites [Courtesy: Premium Aerotec GmbH]

In order to directly manufacture CFRP toolings, special tooling materials, such as HexTOOL®, are required. In contrast to many other prepreg toolings, these materials do not consist of fabrics, but are delivered in the form of quasi-isotropic mats made of short fiber cuts with prepreg resin. Because of their extreme hardness and internal characteristics, it is possible to achieve surface roughnesses (R_a= 1 to 1.5) comparable to those of metals by milling and polishing. Milling of these materials requires considerable experience as well as special milling tools in order to prevent surface defects caused by fiber breakage or matrix cracks.

The major advantage of these materials is the fact that the tooling can be modified to reflect retroactive contour changes by laying up additional layers of the tooling material and milling them to reflect the changed contour. The materials also allow for surface repairs. Neither repair nor retroactive contour adaptation is typically possible with metal tools.

Figure 6.16
CFRP tooling made from HexTOOL® prepreg [Courtesy: Premium Aerotec GmbH]

6.3.2 Indirect Manufacturing of Forming Tools

For the indirect manufacturing of tools, the CAD model data are first used to mill a master mold from a suitable material. This master mold is then used in a second step to form and manufacture the actual tooling. This approach is often used for CFRP and galvano tools. A disadvantage compared to directly manufactured tools is the fact that the tolerances (e.g., with respect to contour accuracy) of the individual manufacturing steps will accrue.

The master in Figure 6.17 is a tooling made from Ebalta material (epoxy tooling block) with a support structure made from wood. The Ebalta mold is floating on the support structure, i.e., it is not permanently connected, in order to avoid stresses caused by differences in expansion coefficients. The surface of this mold represents the contour of the tooling to be manufactured from CFRP. Here, the CFRP tooling is used for the manufacture of a door pressure pad in the door surround structure of an airplane fuselage. The tooling is necessary in order to ensure contour accuracy and to provide the correct interface between fuselage structure and the reinforcing structures at the door cut-out.

Here, the contour accuracy of the tooling is the sum of the manufacturing accuracy of the mold made from Ebalta (milling and contour grinding and/or polishing) and the accuracy of the CFRP tooling production. In the shown example, the final contour accuracy was approx. $\pm$ 0.7 mm.

Manufacture of CFK Tools

There are different approaches to the manufacture of tools made from fiber reinforced plastics for the curing of prepreg components. Often, master molds made from different materials, such as metal, wood, or plastics, are utilized. First, a mirror image of the contour of final the component surface is milled from the tooling material, sealed if necessary, polished, and release agent is applied. The master is then used to laminate the fiber reinforced plastic tooling, utilizing glass- or carbon-fiber fabrics and mold making resins or specific tooling prepregs (Table 6.2).

Figure 6.17
Master for the manufacture of a CFRP tooling [Courtesy: Premium Aerotec GmbH]

The advantage of this approach is the fact that once a master has been produced, it is possible to repeatedly laminate new fiber reinforced plastic tools. Analogous to metal molds, the created geometry is stabilized by a stiff sub-structure, which facilitates safe handling and the mounting of vacuum lines and hoisting points. In order to avoid differences in thermal expansion behavior between mold and stabilizing structure, CFRP structures are often used to reinforce the mold. Depending on surface finish (peak-to-valley height) of the master mold, it is possible to form highly smooth components from CFRP toolings (Figure 6.7)

Manufacture of Nickel-Galvano Molds

While CFRP toolings can be manufactured by "simply" laminating over the master, the path to nickel-galvano tools is more complex. Here, first a master is milled from, e.g., an epoxy tooling block whose contour matches the contour of the final component. Then Ni36 is deposited on the surface of this master (via an electrolytic bath) until the tooling reaches the desired wall thickness. After demolding, the tool surface is ground and polished (mirror finish). In order to stabilize the mold, an egg carton structure made from steel or Ni36 is utilized, similar to other metal toolings. Nickel-galvano toolings (NGTs) are used to manufacture composite components with Class-A surfaces.

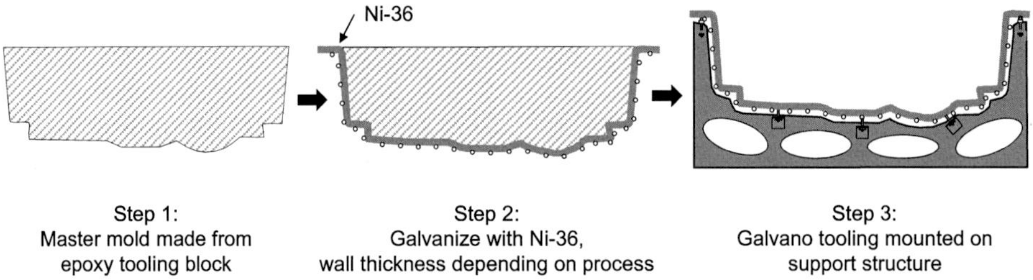

Figure 6.18 Manufacturing sequence for nickel-galvano tooling

Figure 6.19 Nickel-galvano tooling for aircraft components [Courtesy: Premium Aerotec GmbH]

The contour accuracy of NGTs can be adjusted via threaded connections behind the sheet. In Figure 6.19, the adjustment mechanism can be seen in the upper left part. The cover sheet is so "thin" that its contour can be adjusted even after the set-up of the tooling. It allows to meet contour accuracy requirements of $\pm$ 0.2 to $\pm$ 0.3 mm.

6.4 Tooling Concepts and Examples

To bring prepreg materials into a desired form, it often requires not only a mold and curing tool, but also a tooling concept. This is often the case for components with demanding geometries, e.g., multiple-curvature surfaces with reinforcing elements. Frequently, these are multiple-step component generation processes in which each step or individual component requires a specific tooling. In a final step, the individual sub-components, e.g., skin segments and stringers, are assembled and cured. In addition to the forming and curing operations for prepreg components, requirements in terms of surface quality as well accuracy of laminate thickness and radii must be met. Here, it is sometimes difficult to evaluate in advance whether the selected tooling concept will be successful because tooling, material, and process are influencing each other (Chapter 8).

In the following, we will describe a variety of common tooling concepts and their applications. Detailed examples will explain the design of such prepreg processing tools as well as their implementation in the process chain.

6.4.1 Hot Forming Tools

When manufacturing contoured components, such as stringers, it is possible to directly lay up the prepreg on the curing tooling and to form each individual layer. To simplify this process and make it more effective, the prepreg can be processed first into a flat (two-dimensional) stack via ATL or AFP (Section 4.4). Subsequently, this stack is transformed from a two-dimensional preform into the desired three-dimensional contour on a hot forming tooling under heat and pressure. The forming process typically takes place in a solarium (Section 4.7), where the material is first heated and then formed on the tooling. The tooling may be further utilized, e.g., to form a T-profile (commonly used in the aircraft industry) from two L-shaped preforms (Figure 6.22).

For smaller hot forming tools, materials such as wood or epoxy are suitable. For larger tool dimensions, metals or CFRP are better suited because of the required intrinsic stability (dimensional stability). Non-metallic materials offer the advantage of lower thermal capacity that ensures that the tool heats very little, even when used consecutively. This in turn eases the release of the finished preform from the tooling surface, because the tack of the prepreg material, and thus the adhesion to a surface, increases with increasing temperature. Use of a release film facilitates the separation of component and mold. The release film may have to be rather flexible (elastic) to prevent the formation of wrinkles between tooling surface and prepreg stack.

Depending on component contour it is possible to create undercuts in the preform so that it can be directly transferred into the curing tool.

Figure 6.20 shows a relatively small hot form tooling for the manufacture of swaged ribs in a negative form. Here, the prepreg stack to be formed is positioned on the wood block and over the milled swaged hole and then formed via the single-diaphragm method (Section 4.6).

Figure 6.20
Hot form tooling for the manufacture of swaged ribs

Figure 6.21 shows large hot forming tools for the manufacture of stringers. Figure 6.22 shows a manufacturing sequence for the production of T-stringers via hot forming. Here, a flat stack is formed on the tooling and cut. Subsequently, the cores are turned and pressed so that a T-profile is created.

Figure 6.21
Large hot forming tool for the manufacture of stringers [Courtesy: Premium Aerotec GmbH]

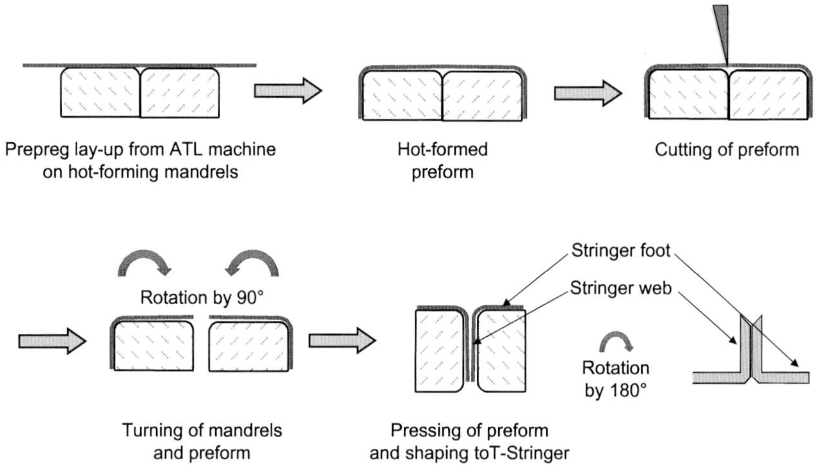

Figure 6.22 Manufacturing sequence for T-stringers

6.4.2 Curing Molds

Molds for the curing of prepreg components can be classified in different categories: open molds and closed molds (Figure 6.23). In open molds, the bottom surface consists of a hard forming tool (e.g., from metal or CFRP), and the top surface is covered by a flexible vacuum set-up with a disposable membrane (film bag), which is sealed (vacuum tight) to the bottom mold at the sides with the help of tacky tape. Instead of the film bag, the top of closed molds is also a hard tooling.

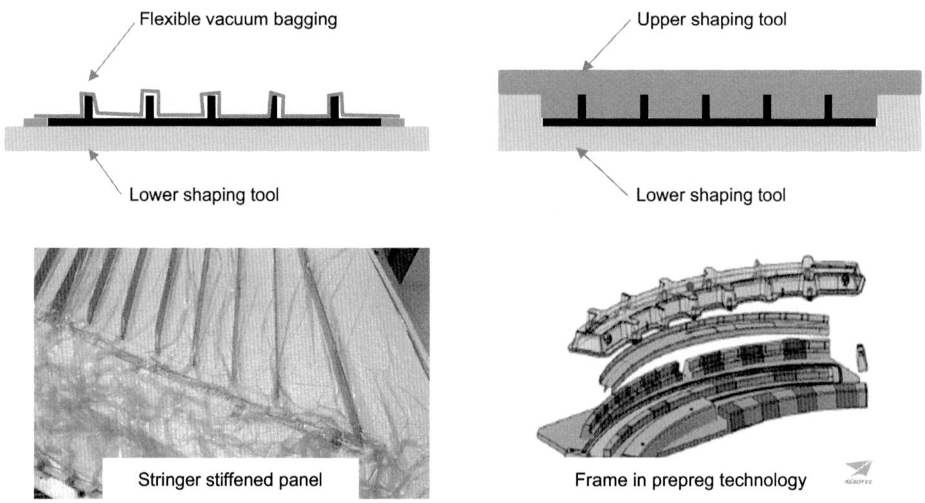

Figure 6.23 Schematics of open (left) and closed (right) mold

Other forming elements are often used, such as curing mandrels, where we differentiate between rigid and flexible core systems. The schematic in Figure 6.24 shows a potential configuration of tooling (curing core and bottom tooling), component, and film membrane. They are used to, for example, position uncured stringers on an uncured skin segment with narrow form and positioning tolerances and subsequently curing them together.

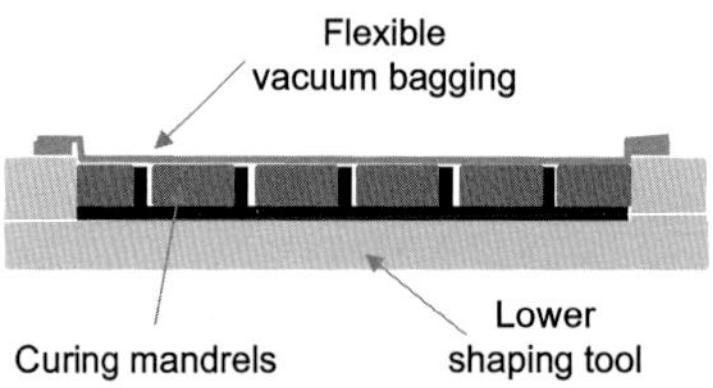

Figure 6.24
Curing mandrels in open mold principle processes

Table 6.5 provides an overview of the advantages and disadvantages of open and closed molds. Curing of the component depends on the complexity of the component and the sequence of the production process, among other factors. Therefore, it is not possible to make general statements as to which mold should be used for what kind of application.

Table 6.5 Overview of Advantages and Disadvantages of Open and Closed Mold Systems

	Advantages	Disadvantages
One-sided closed: open mold	▪ Almost no constraints in designing the inner mold line	▪ Undefined surface structure on bagging side ▪ High amount of bagging material (no re-use)
Two-sided closed: closed mold	▪ Very well defined and smooth surfaces (whole composite part surface), well prepared, e. g., for subsequent painting or assembly	▪ High capital investment ▪ Very heavy tools ▪ Tooling size limited ▪ Perfect match between CTE of composite part and tooling necessary

Table 6.6 provides an overview of the materials used for curing mandrels as well of the advantages and disadvantages of curing mandrels. Open mold applications utilize rigid mandrels made, e. g., from steel or aluminum, as well as flexible mandrels made from silicone or natural rubber. The selection of mandrel material depends on the processing concept as well as on the component design. Mandrels such as those shown in Figure 6.24 are commonly used because they are easily demolded. The service life of the mandrel material is of particular importance for high production runs.

Table 6.6 Overview of Rigid and Flexible Mandrel Technologies

	Rigid mandrel technique	Flexible mandrel technique
Materials	▪ Steel, aluminum, or CFRP mandrels ▪ Airpad mats, rigid foams ▪ Aquacore™, sand-filled tubes	▪ Polyamide/polyimide film ▪ Silicone and natural rubber ▪ Airpad mats
Advantages	▪ Very smooth/polished surface ▪ Exact angles and contour accuracy ▪ Nearly unlimited durability of curing mandrel	▪ Good laminate quality ▪ Good contour accuracy ▪ No constraining force caused by thermal expansion ▪ Low price/fast/easy to manufacture
Disadvantages	▪ Constraining force caused by thermal expansion, esp. with metal mandrels ▪ Not possible to uniformly distribute autoclave pressure, laminates prone to porosity ▪ Heavy and complex construction	▪ Very limited service life ▪ Limited possibilities to provide tight tolerances ▪ Constant production of flexible mandrels necessary

Suitable mandrel materials for the molding of hollow parts, such as Omega profiles and pipes, are rigid materials such as Aquacore™ (Figure 6.25) or sand-filled tubes. Aquacore™ is a heat-resistant and water soluble material for the manufacture of so-called lost cores. It is available in the form of wet powders and can be brought into the desired form manually or by press-forming. The mandrel material can be easily washed out ("demolded") with tap water after curing. In general, the material can be reclaimed by drying.

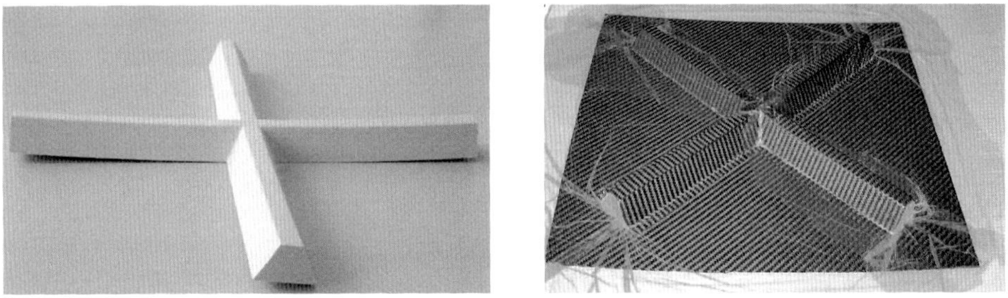

Figure 6.25 Aquacore™ as filler material [Courtesy: Aero Consultants AG]

Flexible mandrels can be made from film or rubber tubes that are inflated by compressed air (or autoclave pressure). Figure 6.26 shows various hollow profiles that were manufactured via tube technology and an external mold.

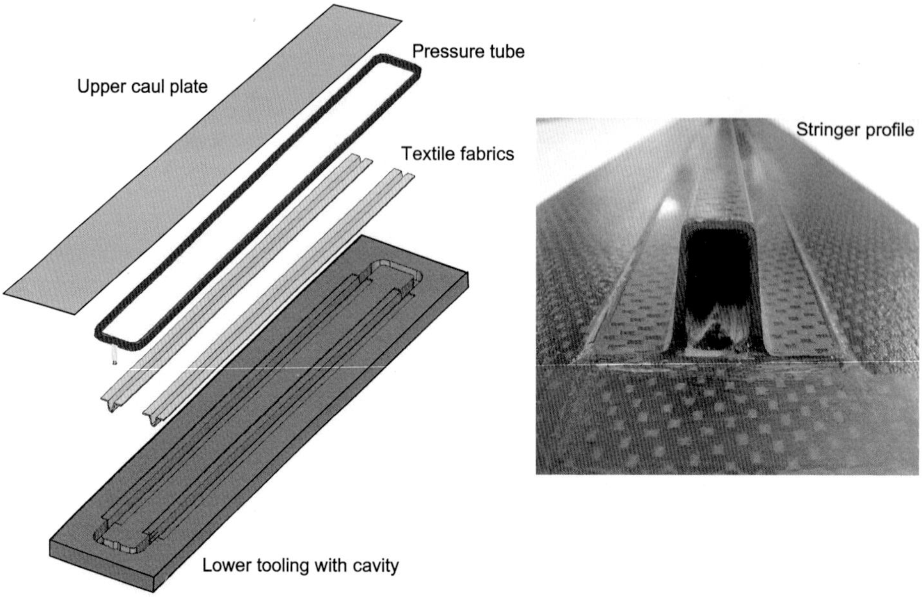

Figure 6.26 Tooling for manufacture of hollow profiles

A special variation of the rigid core technology is the utilization of rigid foams as the forming tool, e.g., for the manufacture of Omega-stringers during their simultaneous integration on a component. This is another case of the "lost-core" approach, because the foam core remains in the component.

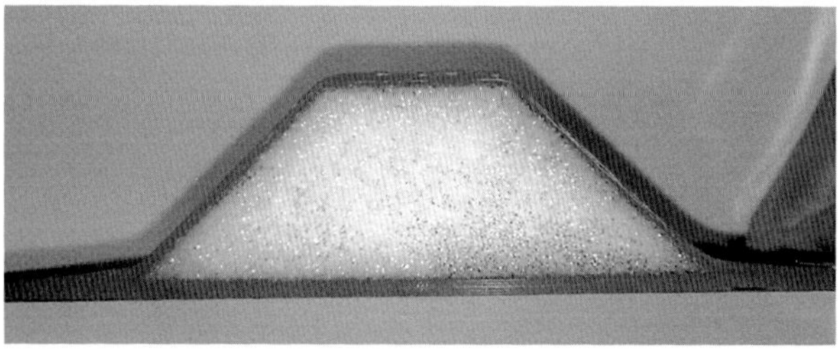

Figure 6.27 Foam core used as forming tool for Omega profile

The Airpad mat represents a peculiarity as material for cores, because it can be used for both rigid core and flexible core applications. Airpad is a non-cured, silicone-free rubber that can be used for the manufacture of pressure plates and of flexible cores. The mat facilitates increased pressure or a more uniform pressure distribution on the component during autoclave curing. This allows for almost identical surface quality on the vacuum bag side of the component and on the tooling side. Airpad is resistant to approx. the same temperatures as silicone; however, it does not cause silicone contamination. Airpad mats can be reinforced by adhering them to tooling prepregs, thus providing them with dimensional stability. This approach allows for the local adaptation of flexibility and stiffness.

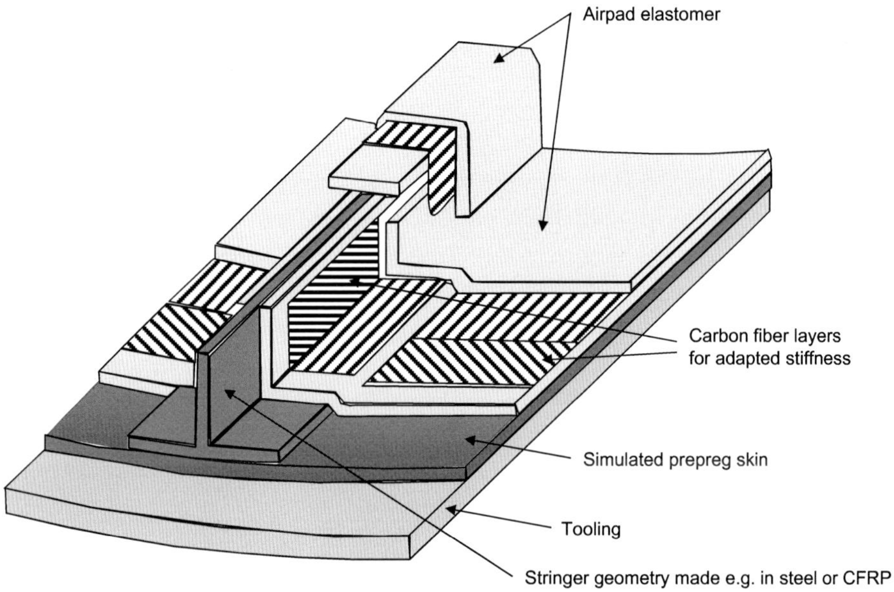

Figure 6.28 Design for manufacturing with reinforced Airpad mat

6.4.2.1 Open Mold Curing Tools

As discussed earlier, the upper side of open curing molds are always set up with flexible, adjustable vacuum equipment including a film bag made of a disposable membrane (Figure 6.29). This technology offers a variety of options to integrate additional elements on the top face of the component, such as stringers on large-area components. Here, we distinguish between three commonly used integration variations with respect to the curing states of stringers and skin segment:

a) rigid/soft (stringer cured, skin segment uncured)

b) soft/rigid (stringer uncured, skin segment cured)

c) soft/soft (both mating parts uncured)

As equivalents to the terms "rigid" and "soft", the terms "cured" (= rigid) and "wet" (= soft) are also used.

A fourth variation is the case in which both mating parts are already cured, which corresponds to subsequent adhesive bonding. The variations a) through c) will be explained in more detail in Chapter 8.

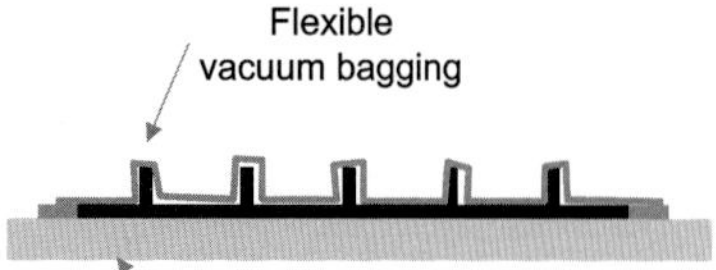

Figure 6.29 Principle of open mold curing tool

6.4.2.2 Closed Mold Curing Tools

In contrast to the open mold variations, which always operate with a flexible vacuum membrane to counter the autoclave pressure, closed mold curing tools work with two forming tool halves (Figure 6.30). The tool halves are designed to seal each other, making a vacuum membrane redundant. Because of the high costs for upper and lower tool halves, this mold concept is typically used only for components with smaller dimensions. Closed mold toolings are used considerably less often for the curing of prepregs, because the material thickness of the prepreg and the cavity depth in the tool have to be very closely aligned. If these two parameters do not match, the mold either does not close (stack too thick), or the stack is not under enough pressure (stack too thin).

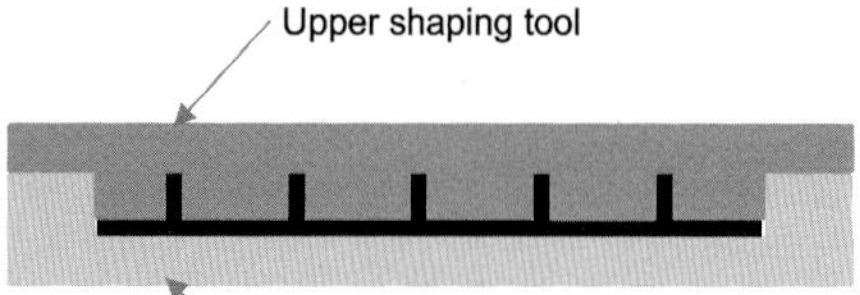

Figure 6.30 Principle of closed curing mold

A combination of closed mold and resin transfer molding (RTM) for prepreg curing has been under investigation for several years (Chapter 4). Here, the pressure necessary for the compaction of the prepreg is created by a free-flowing RTM resin.

6.4.3 Release Agents and Cleaning

For prepreg curing operations, the mold surface always has to be treated with a release agent to make sure that component and surface of the curing tooling do not adhere to each other. Typically, the release agent is reapplied prior to each curing cycle to ensure successful release. Over time and with repeated application of release agent, a layer consisting of new release agent as well as residues of degraded release agent and epoxy resin from the prepreg will build up on the mold surface. This typically leads to a loss in surface quality on the composite component, because the contamination is partially transferred from the mold surface to the laminate. Therefore, these residues have to be removed regularly from the mold surface.

Cleaning is generally carried out by wet grinding of the surface, followed by polishing and reapplication of release agent. Other options are stripping of the surface contamination via laser or dry ice blasting.

6.5 Additional Information

[1] G. Niederstadt. *Ökonomischer und ökologischer Leichtbau mit faserverstärkten Polymeren.* Renningen: Expert-Verlag, 1997.

[2] M. Kühn, M Hanke, S. Malzahn, T. Wurl. *Elastomere Formwerkzeuge in der Faserverbundtechnologie – Möglichkeiten und Grenzen* (Paper). Braunschweig: Deutscher Luft- und Raumfahrtkongress, 2010.

[3] P. Ermanni. *Composites Technologien.* ETH-course. Zürich: ETH 2007, August.

[4] H. Apmann. *Vorteile durch CFK und Leichtbau.* Company presentation Premium Aerotec GmbH. 2013.

[5] Technical data sheet, HexTOOL® M61, Hexcel Composites, 2014.

[6] Technical data sheet, HexTOOL® M81, Hexcel Composites, 2014.

[7] Technical data sheet, DURATOOL 5270-2, Cytec Engineered Materials, 2013.

[8] Technical brochure, HexTOOL Fabrics UserGuide, Hexcel Composites, 2013.

[9] Technical data sheet, Beta Prepreg, AIRTECH Europe Sarl, 2013.

[10] OBO product brochure: Blomberger Holzindustrie B. Hausmann GmbH & Co. KG, issue April 2012.

7 Testing of Prepregs

Johannes Krämer

Fiber reinforced plastic materials (FRP) exhibit anisotropic and therefore very complex material behavior; as a result, the simulation of their behavior in support of component design and dimensioning is also highly complex. This material class also requires specific testing procedures that differ from conventional tests for standard plastics.

The properties of FRP are mainly determined by fiber type, matrix material, and fiber volume content.

The effect of the manufacturing process on the final properties has to be taken into account when determining mechanical and thermomechanical properties of FRPs. The curing cycle affects the properties of both matrix and interphase, and micro-defects created during the manufacturing process may later lead to component failure.

This is of particular relevance for prepregs, because here the manufacturing of a component also includes the processing of a semi-finished product. Therefore, this chapter will focus on the characterization of semi-finished products, i.e., the uncured prepregs. On the one hand, these characterizations are relevant for the manufacturer in terms of quality control; on the other hand, they are also tools for material testing on arrival for the subsequent user.

Other topics relevant with respect to FRPs, such as non-destructive testing or (thermo-)mechanical testing, will not be covered in detail here; instead, reference to the existing literature will have to suffice [1–3].

From the point of view of the processor, the two most important properties of a prepreg are drapability and tack. However, there is a large number of testing procedures and test variations, of which we will describe a selection in more detail. Where available, the respective standards will be referenced.

7.1 Characterization of the Uncured Prepreg

7.1.1 Prepreg Areal Weight, Resin Content, and Fiber Areal Weight

The prepreg areal weight (PAW) is gravimetrically determined by weighing a prepreg sample of defined size. In industrial practice, cut circles with an area of 100 cm^2 are common. In order to determine the prepreg areal weight in g/m^2, the mass of this disk only has to be multiplied by a factor of 100.

This sample can simultaneously be used to determine the resin content (RC) of the prepreg in wt.%. For this purpose, a suitable solvent is used to wash the resin from the matrix of the weighed prepreg sample. The remaining fibers are separated from the solvent, dried, and weighed. The weight differential represents the resin content. The fiber weight, with respect to the sample area, represents the fiber areal weight (FAW).

Standard references:

- ASTM D 3171: Standard Test Methods for Constituent Content of Composite Materials (2011)
- ASTM D 3529: Standard Test Method for Matrix Solids Content and Matrix Content of Composite Prepreg (2010)
- ASTM D3776: Standard Test Methods for Mass Per Unit Area (Weight) of Fabric (2013)
- DIN EN 2329: Aerospace series; textile glass fiber preimpregnates; test method for the determination of mass per unit area (1993-04)
- DIN EN 2559: Aerospace series – Carbon fiber preimpregnates – Determination of the resin and fiber content and the mass of fiber per unit area (1997-05)

7.1.2 Resin Flow

The resin flow, as a measure of the flowability of the resin, is an important parameter for the processing of prepreg systems. When determining whether there still is a sufficient flow of the matrix during the later stages of the curing cycle, the test also provides information regarding the degree of prepreg aging. Thus, determining resin flow also serves as a measure of quality control.

The resin flow is determined by measuring the mass loss of samples after curing under set conditions. Here, the following standards are relevant:

- ASTM D 3531-99: Standard Test Method for Resin Flow of Carbon Fiber-Epoxy Prepreg (2011)
- DIN EN 2332: Aerospace series; textile glass fiber preimpregnates; test method for the determination of the resin flow (1993-04)
- DIN EN 2560: Aerospace series – Carbon fiber preimpregnates – Determination of the resin flow (1998-08)
- IPC-TM-650 No. 2.3.17: Resin Flow Percent of Prepreg (08/1997)

The standards differ in geometry, set-up, and parameters; however, in principle the methods are the same. In the following we will detail the determination of the resin flow according to DIN EN 2560:

The prepreg is brought to room temperature and subsequently stored in a conditioned atmosphere for at least 2 h. The measurement is to take place within the following 6 h. At least three square prepreg samples (edge length 100 mm) are cut according to a cutting pattern, stacked crosswise, and weighed. The sample is then placed symmetrically between a layer each of perforated polyvinyl fluoride film (PVF film), three layers of glass filament fabric, and three layers of glass filament fabric with polytetrafluoroethylene coating. This lay-up is compressed between two sheets in a temperature-controlled fixture (Figure 7.1) and subsequently cured. For this procedure, temperature, pressure, and curing time are to be selected according to the prepreg performance data sheet.

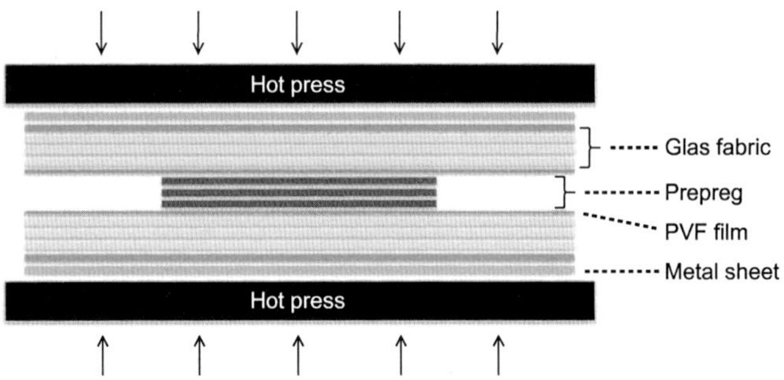

Figure 7.1 Schematic set-up for resin flow measurement

Once cooled, the film set-up is removed. A cuboid and a cylinder are cut from the sample and weighed.

The non-adjusted resin flow (RF) is calculated from at least two samples according to Eq. 7.1 and stated in percent.

$$RF = 1 - \frac{W_2 A_1}{W_1 A_2} \times 100 \tag{7.1}$$

Where:

RF non-adjusted resin flow percent of initial prepreg mass

W_1 sample weight prior to curing

A_1 initial square area of the sample prior to curing

W_2 weight of sample cut after curing

A_2 rectangular and circular areas of the samples cut after curing

The resin flow adjusted for the volatile content (see Section 7.1.5) is calculated according to Eq. 7.2:

$$RF_{\text{adjusted}} = 1 - \frac{100 W_2 A_1}{W_1 \left(100 - VC\right) A_2} \times 100 \tag{7.2}$$

Where:

RF_{adjusted} adjusted resin flow percent of initial prepreg mass

VC percentage of volatiles (see Section 7.1.5)

7.1.3 Tack

Tack describes the ability of prepreg materials to adhere to themselves or to the surface of a mold or tool. This characteristic influences processing: tack has to provide enough stability to prevent disorientation of fibers during processing, while also offering an opportunity for corrections.

A prepregs tack is determined by several factors. The processing parameters, such as temperature, as well as the initial material and its manufacturing and storage conditions, play a decisive role. Here, an important characteristic is the tack life. It defines the time for which a prepreg may be stored at room temperature while retaining sufficient tack to facilitate processing (Chapter 3).

There is no standardized scale for the degree of tack established in industrial practice because of the wide variety of materials and other potential influences. There are also no standardized, universal testing methods, but rather only application-specific, adapted methods.

The probe tack test is one such method to determine the tack of a prepreg. Here, a temperature-controlled metal probe is first brought into contact with the prepreg and then removed while the force and/or energy of separation is measured [4, 5]. Important parameters influencing the results of these measurements are the probe temperature, contact force, contact time, and rate of debonding.

7.1.4 Drapability

The drapability describes the ability of textiles and semi-finished products to deform and adapt to a three-dimenional contour. How well a semi-finished product can be draped depends first and foremost on the textile composition. For example, a plain weave has better drapability than a twill weave. The strong restoring forces exhibited by prepregs with inferior drapability will impede component lay-up. However, if drapability is too high, the risk of losing the orientation of the cuts increases because they are too easily deformed.

In order to determine the drapability of a prepreg, the material is simply bent over a cylindrical mandrel (e.g., with a diameter of 3 mm) and subsequently evaluated with a magnifying glass to detect fiber fracture.

7.1.5 Volatiles

For prepregs manufactured via the solution route, the volatile content (VC) of the uncured prepreg is stated. It is measured as the percentage weight loss after a set (temperature/time) exposure to elevated temperature, as defined by the respective material standard.

The prepreg is brought to room temperature and subsequently stored at conditioned atmosphere for at least 2 h. Several square prepreg samples (e.g., 100 mm × 100 mm) are cut and their initial weight is determined in an aluminum tray. After approx. 20 min exposure at 160 °C in an air circulating oven, the sample is cooled to room temperature in a desiccator and weighed again. Depending on testing procedure, this step may be performed with the help of a tray or a hanging device in the oven. The weight differential, normalized against the initial prepreg weight, is used to determine the volatile content according to Eq. 7.3.

$$VC = \frac{W_1 - W_3}{W_1 - W_2} \times 100 \tag{7.3}$$

Where:

VC percentage of volatile contents in the initial mass

W_1 initial sample weight (including aluminum tray)

W_2 weight of aluminum tray

W_3 sample weight after heating (including aluminum tray)

Typical values for the VC of solvent process prepregs range between 2 % and 8 %. For hot-melt prepregs the VC is practically zero and therefore typically not stated.

Standard references:

- ASTM D 3530: Standard Test Method for Volatiles Content of Composite Material Prepreg (1997)
- DIN EN 2330: Aerospace series; textile glass fiber preimpregnates; test method for the determination of the content of volatile matter (1993-04)
- DIN EN 2558: Aerospace series – Carbon fiber preimpregnates – Determination of the volatile content (1997-03)
- IPC-TM-650 No. 2.3.19: Volatile Content of Prepreg (12/1994)

7.1.6 Viscosity

In general, the curing of reactive resins can be described with the help of TTT-diagrams (time-transition-temperature) [6]. The chemical reactions during the cross-linking of the matrix turn the viscous resins into solids. In other words, the viscosity increases as the curing process advances. Knowledge of the progression of the viscosity of a prepreg material is valuable for the determination of the appropriate processing window when curing it into final components. It is also an important parameter for the manufacture of prepregs because the viscosity can be used to control the level of impregnation and thus the entire process (see Chapter 3).

Measuring the dynamic viscosity provides information on the behavior of the matrix during heating in the autoclave. The faster the system heats up, the more the absolute viscosity value reduces (caused by the increase in temperature); on the other hand, the increase in temperature also initiates a faster viscosity increase as a result of the cross-linking reaction (Figure 7.2). This behavior indicates at which point in time the autoclave pressure should be built up (see Chapter 5). If viscosity measurements are continued on the cross-linked sample, additional information can be obtained with regard to the gel point, when the state of complete state is reached, and when cooling can be started.

Isothermal viscosity measurements (Figure 7.3) provide the processing parameters for the selected prepreg material: on the one hand the temperature required to set the absolute viscosity (begin of measurement). On the other hand, it allows for an estimate of the respective processing window, i.e., for how long the viscosity will remain within a certain range (see Chapter 3).

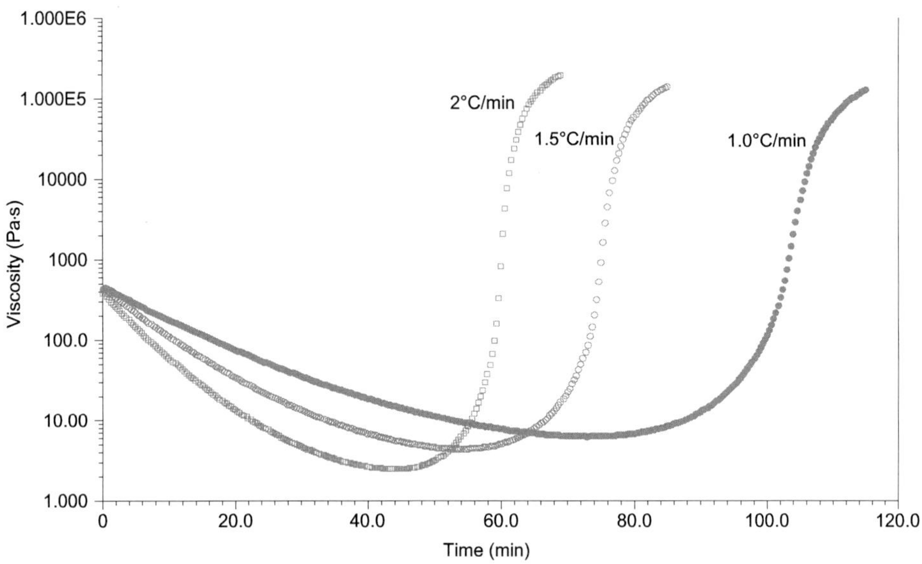

Figure 7.2 Dynamic viscosity of an epoxy resin system

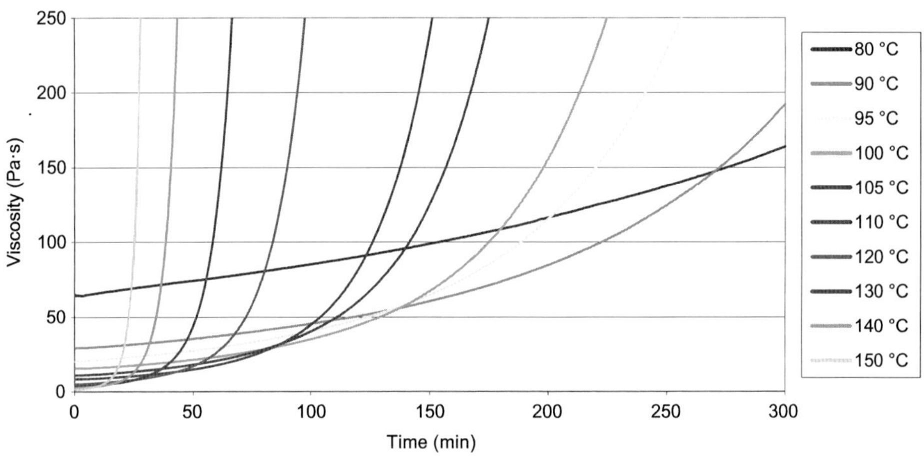

Figure 7.3 Isothermal viscosity measurements of an epoxy resin system

Standard references:

- ASTM D 4473: Standard Test Method for Plastics: Dynamic Mechanical Properties: Cure Behavior (2008)
- DIN 16945: Reaction resins, reactants and reaction molding materials; test methods (1989-03)

7.1.7 Water Pickup Test (WPU Test)

The WPU test essentially serves as an inspection tool to determine the level of impregnation in-line during the manufacture of prepreg materials. The results are required in order to adapt the process parameters to the material specifications. To determine the WPU, a sample of defined size is cut and weighed. The sample is then vertically submerged in distilled water at a defined depth and time. While the sample is submerged, capillary forces will absorb water between the fibers. Depending on the portion of non-impregnated fibers, the sample will absorb more or less water. After the set test time, water adhering to the sample exterior is dried and the sample is weighed again. From the difference in sample weight the percent WPU can be calculated. This test can be performed only with fresh material during the prepreg manufacturing operation because the subsequent storage (freezing) and handling will affect the measurement results. Depending on prepreg material and required level of impregnation, the results typically exhibit large variations. However, they provide an indication for the manufacturer whether the prepreg line is running within specifications. The measurements in general do not provide quality criteria within a specification.

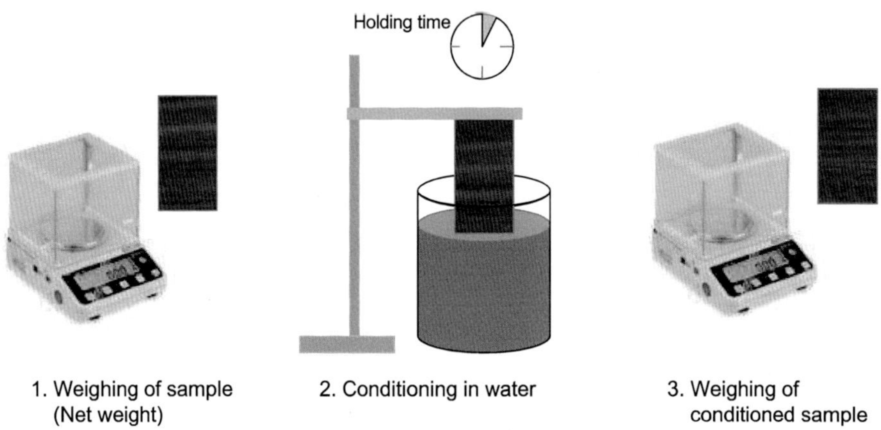

Calculation: $WPU = \frac{\text{(Weight of conditioned sample – Net weight) [g]}}{\text{Net weight [g]}} \times 100\ \%$

Figure 7.4 Sequence of WPU testing

7.1.8 Microscopy Cut Edge Test

In addition to the WPU test, there is an optical method for the determination of the level of impregnation. Here, the cut edge of a prepreg sample is investigated and a specified length is photographed (black and white images) [10]. Depending on the material, distinct differences in the level of impregnation can be distinguished with the naked eye. An image evaluation software calculates the area of non-impregnated regions on the basis of their gray-scale values and shows them with the help of a so-called false-color representation. The ratio of different colored areas determines the level of impregnation. Similar to the WPU test, this method cannot be used to make universal statements, but it rather indicates a trend in the level of impregnation based on several measurements and materials. Sample preparation, measurement, and interpretation require experience and prudence, because many determining factors have not yet been investigated sufficiently.

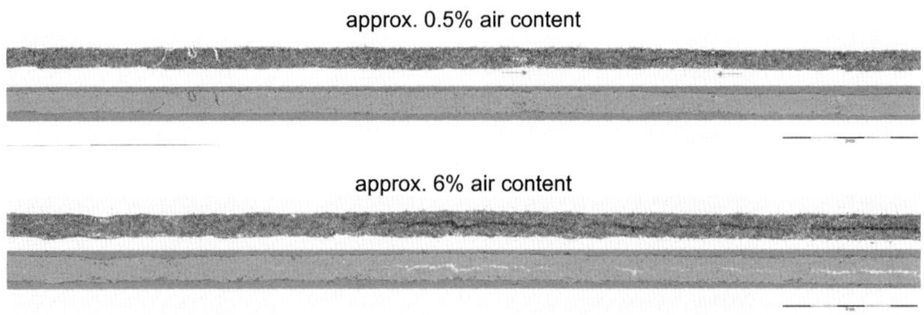

Figure 7.5 Determining the level of impregnation via optical measurements

■ 7.2 Characterization of the Cured Prepreg

7.2.1 Fiber Volume Content

While suppliers of semi-finished products typically provide information regarding the matrix/fiber ratio as a weight ratio (Section 7.1.1), component design and quality management require it in terms of fiber volume content (FVC). Both values differ because of the differences in material densities of matrix and fibers. The fiber volume content can be determined theoretically either as a calculation with respect to laminate thickness, via the ash content, via wet-chemical dissolution, or optically through micrographs. The determination of the FVC via laminate thickness measurements or via optical methods is relatively inaccurate and does not provide information regarding porosity; the other methods are time-consuming.

For testing via the ash content, a sample is weighed together with its ceramic carrier and subsequently the polymer matrix is pyrolyzed in a microwave oven. This method can be used only for glass fiber reinforced plastics, because the high temperatures required would cause oxidation of the carbon fibers.

Figure 7.6 shows the testing sequence of wet-chemical fiber volume determinations. After weighing, the fibers are extracted from the matrix by a wet-chemical process utilizing nitric acid. Subsequently, the fibers are washed, dried, and weighed.

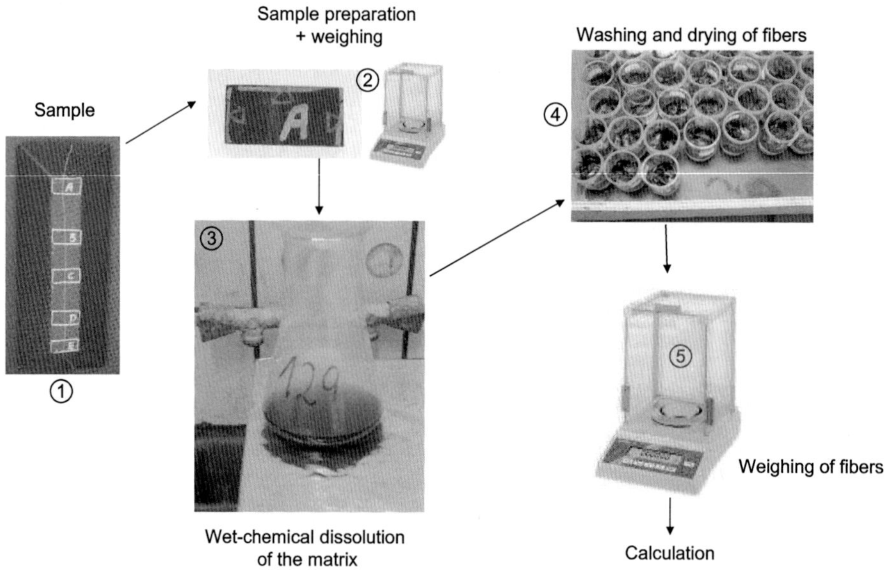

Figure 7.6 Sequence of wet-chemical fiber volume content determination

In both cases, the fiber volume content V_f is calculated from the densities and weights of both fiber and matrix according to Eq. 7.4 and stated in percent.

$$V_f = \frac{\rho_m W_f}{\rho_f W_m + \rho_m W_f} \times 100 \tag{7.4}$$

Where:

ρ_m matrix density

ρ_f fiber density

W_m matrix weight

W_f fiber weight

Alternatively, the fiber volume content can be calculated with the help of microscopy using the area or line method [11].

Standard references:

- DIN EN 2564: Aerospace series – Carbon fiber laminates – Determination of the fiber-, resin- and void contents (1998-08)
- ASTM D3171: Standard Test Methods for Constituent Content of Composite Materials (2011)

7.2.2 Determination of Glass Transition Temperature

Dynamic mechanical thermal analysis (DMTA) is a classical method to determine the glass transition temperature T_g. Here, a small sample is exposed to oscillating torsional or bending loads (mostly sinusoidal at 1 Hz) while the temperature of the sample is continuously increased at a defined heating rate. Figure 7.7 shows the result of this measurement: the characteristics of the storage modulus, the loss modulus, as well as the loss factor (tan δ). It is important to always reference how the T_g was evaluated, because the results will differ depending on the method chosen. Typically, either the maximum of the loss factor, the maximum of the loss modulus, or the tangent method (decrease of the storage modulus) can be used for the evaluation (Figure 7.7).

These measurements also provide information regarding the degree of curing. If the sample is not completely cured, the tan δ exhibits a wider peak and the storage modulus increases again after the initial decrease at the glass transition point.

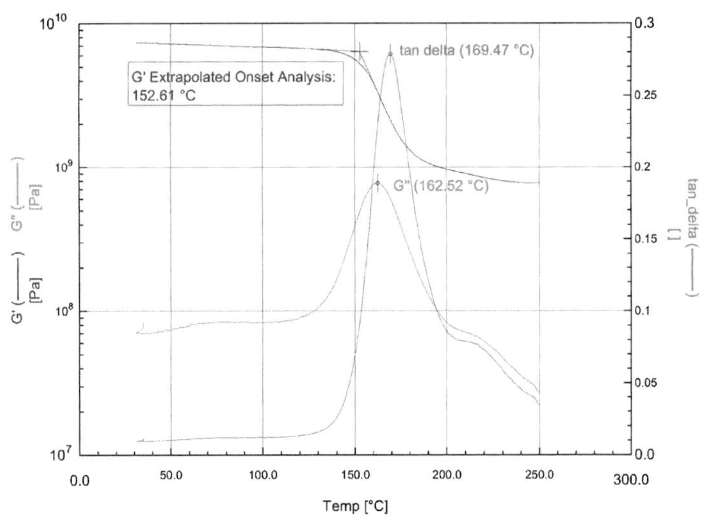

Figure 7.7 DMTA diagram for the determination of the glass transition temperature [Courtesy: T. Neumeyer]

Differential scanning calorimetry (DSC) is used to determine the T_g of an uncured prepreg. This method uses a very small sample (approx. 0.01 g) and a reference sample; both samples are heated at a rate of typically 10 to 20 K/min, and the heat flow is measured. The glass transition can be recognized as a step in the heat flow curve (Figure 7.8). The glass transition temperature of the uncured resin, $T_{g,\,uncured}$, is often used as a measure for quality management. It allows for the determination of whether the prepreg characteristics have changed after extended storage, i.e., how far the chemical reaction already has progressed.

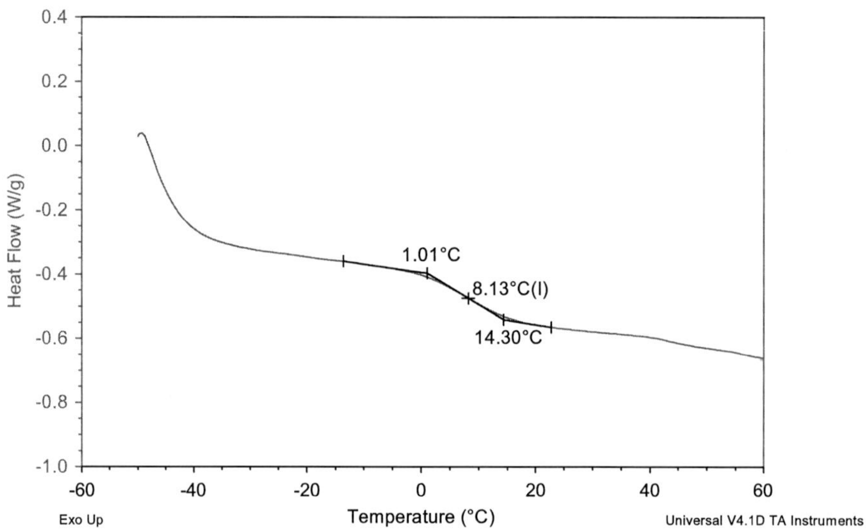

Figure 7.8 DSC diagram for the determination of $T_{g,\,uncured}$

Standard references:

- DIN EN ISO 11357: Plastics – Differential scanning calorimetry (DSC) – Part 2: Determination of glass transition temperature and glass transition step height (2013-02)
- DIN 65583: Aerospace – Fiber reinforced materials – Determination of glass transition of fiber composites under dynamic load (1999-04)

7.2.3 Cured Ply Thickness (CPT)

The cured ply thickness is the theoretical thickness of a cured prepreg ply. This value can be calculated from areal fiber weight, fiber volume content, and fiber density according to Eq. 7.5.

$$CPT = \frac{FAW}{V_f \times \rho_f \times 10} \quad (7.5)$$

Where:

CPT thickness of cured prepreg ply in mm

FAW prepreg areal fiber weight

V_f fiber volume content

ρ_f fiber density

Experimentally, the CPT can be determined as the quotient of component thickness and number of individual plies.

7.3 References

[1] W. Grellmann, S. Seidler. *Polymer Testing.* 2nd ed., Munich: Hanser Publishers, 2013, Chapter 10.

[2] G. W. Ehrenstein. *Faserverbundkunststoffe.* 2nd ed., Munich: Hanser Publishers, 2006, Chapter 6.

[3] G. W. Ehrenstein, G. Riedel, P. Trawiel. *Thermal Analysis of Plastics: Theory and Practice.* 2nd ed., Munich: Hanser Publishers, 2004.

[4] K. J. Ahn, J. C. Seferis, T. Pelton, M. Wilhelm. *Analysis and Characterization of Prepreg Tack.* Polymer Composites. 1992, 13, pp. 197–206.

[5] T. Neumeyer, M. May, J. Krämer, V. Altstädt. *Influence of the Network Structure of DGEBA Based Epoxy Resins on their Tack* (Poster). Berlin: Thermosets, 2011.

[6] M. Wacker, P. Trawiel, G. W. Ehrenstein. *Härtung von Reaktionsharzen Time-Temperature-Transition-Diagramm* (reprint). Erlangen: Department of Plastics Technology, 2007.

[7] Hexcel company brochure: *Prepreg Technologies.* Hexcel Composites, 2012.

[8] *Prepreg Materials Characterization.* MIL-Handbook-17-1E. 2002, Chapter 5, pp. 1–14.

[9] R. J. Crossley, P. J. Schubel, N. A. Warrior. *Experimental Determination and Control of Prepreg Tack for Automated Manufacture.* Plastic, Rubber and Composites. 2011, 40, pp. 363–368.

[10] Patent DE102006033663B4: *Verfahren zur Ermittlung eines charakteristischen Parameter an einer CFK-Probe.* 2012.

[11] W. Grellmann, S. Seidler. *Polymer Testing.* 2nd ed., Munich: Hanser Publishers, 2013, p. 548 f.

8 Design and Production

Hauke Lengsfeld

This chapter will describe the different design concepts that can be used to convert prepreg materials into a component or an assembly. The choice of design concept has far-reaching consequences for the following process sequence and the subsequent assembly of the component, because design, material-specific characteristics, component design, process, etc. are interdependent.

8.1 Design Concepts

When converting prepreg materials into a component using a tooling, different design concepts can be applied. They depend on the geometry of both component and tooling as well as on the prepreg lay-up process selected. The design also depends on whether a particular side or surface of the final component has to meet certain requirements of potential post-processing steps, such as assembly (smoothness).

Independent of the design, we will discuss only open curing molds (see also Chapter 6), i. e., the top surface of the component will be covered by a flexible, adaptable vacuum equipment consisting of a film bag and a disposable membrane.

8.1.1 Positive/Negative Design

In general, we distinguish between two design concepts with regard to the tooling contour:

- Positive design
- Negative design

In a *positive mold* the prepreg is laid up on the tooling to form a concave shape (Figure 8.1, left). The interior surface of the component will be smooth and able to meet strict tolerance demands, while the outside surface will have an undefined structure. This mold design is suitable in particular for long, narrow components, such as double-T stringers and spars (Figure 8.1, right). This variation is also often used for hot-forming molds.

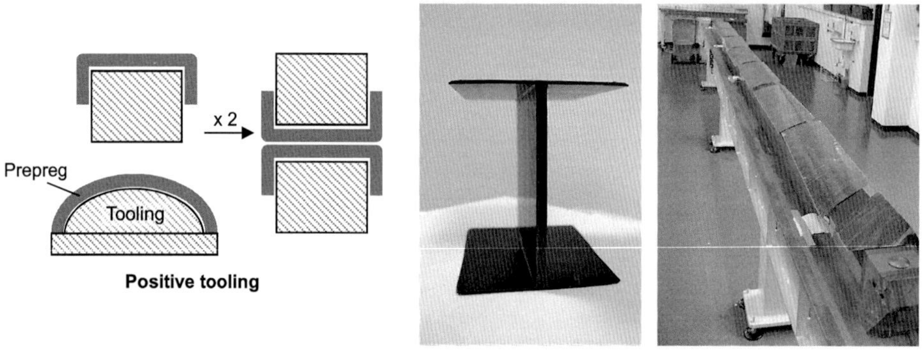

Figure 8.1 Lay-up of prepreg on positive mold and example of component

An advantage of this geometry is that it is typically easier to lay up the material over a positive mold than into a mold cavity. However, this also depends on the processing characteristics of the material and on the laminate structure (e.g., fiber stiffness, tack, and layer orientation). This design is suitable in particular when the coefficient of thermal expansion/cooling shrinkage of the mold is significantly higher than that of the fiber reinforced component after curing. Here, the mold will "shrink free" of the component, facilitating demolding.

In a *negative mold,* the prepreg material is laid up in a mold where it is formed into a convex shape (Figure 8.2, left). Here, the material is draped into a mold cavity. This design is suitable when the outer surface of the component has to meet tolerance demands, e.g., for subsequent assembly purposes. However, the components should have sufficient width to allow for easy, defect-free draping in the mold cavity. The design is suitable for long, wide components, such as pressure calottes and C-spars (Figure 8.2, right and Figure 8.3). Another advantage of this design is the fact that additional geometries, such as stringers, can be integrated on the upper surface of the component.

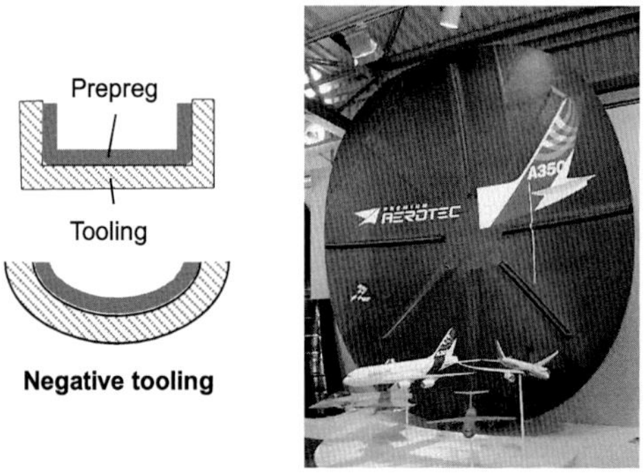

Figure 8.2
Lay-up of prepreg in negative mold and example of component

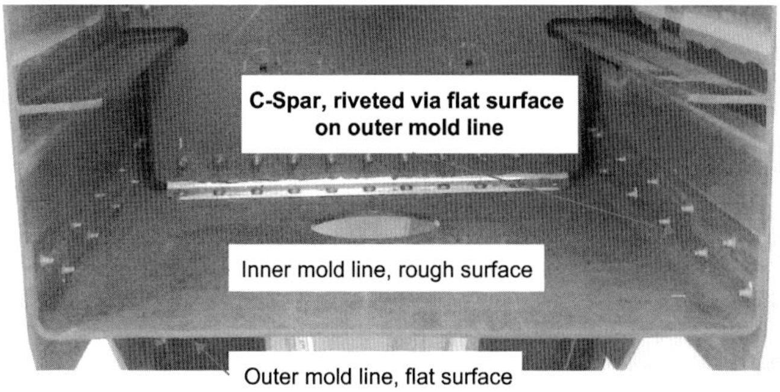

Figure 8.3 C-spar (negative design) once assembled

With respect to potential integration options, negative molds in particular can be further distinguished.

8.1.2 Integral/Differential Design

A design frequently used in the aerospace industry (but also in other composite applications) is based on stringer-reinforced segments (skins or panels). These skin panels can be flat, but they can also have three-dimensional curvatures, such as wing fairings.

Figure 8.4 Stringer-reinforced panel made of CFRP [Courtesy: Premium Aerotec GmbH]

These composite parts are often assembled from a variety of individual elements, such as skin panel, stringers, and spars, in order to achieve minimal weight. The integration of additional individual elements into a final component can be distinguished in two more monolithic design concepts:

- Integral design
- Differential design

Sandwich design could be considered a third variation; however, it requires the use of an additional material and therefore differs in principle from the monolithic designs.

Integral Design

Integral design is a single-step integration process, i.e., all individual elements such as skin panel, stringers, and spars, are assembled in one step and then cured to form the final component (Figure 8.5). This requires that the individual elements are still in an uncured state. Therefore, this method is also called a *soft/soft* variation (see Section 6.4.2.1). Here, "*soft*" describes the state of the individual elements, i.e., the uncured state.

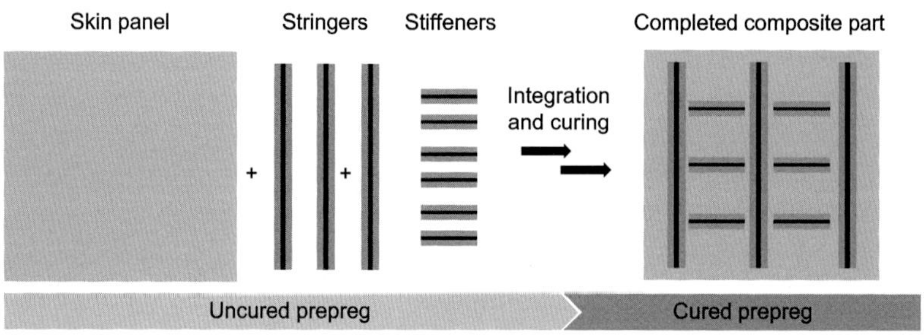

Figure 8.5 Single-step integration process for uncured elements

The advantage of this design is the fact that highly integrated components, i.e., components that consist of many elements, can be manufactured maintaining lightweight construction. A variety of other functions may also be integrated, such as threaded bushings for assembly purposes. The single-step integration of stringers and ribs in a skin panel allows for a reduction in individual elements that would have to be assembled post-curing.

A disadvantage of this technology is the fact that it requires very high tooling efforts (see also Section 8.1.2). The mold and all its individual parts are very closely aligned and even the smallest change, e.g., in geometry or dimensional stability, may lead to defective components. At the same time, the high level of integration also means that the risk of defects, such as voids, in the final component will increase, because none of the individual components can be tested prior to curing.

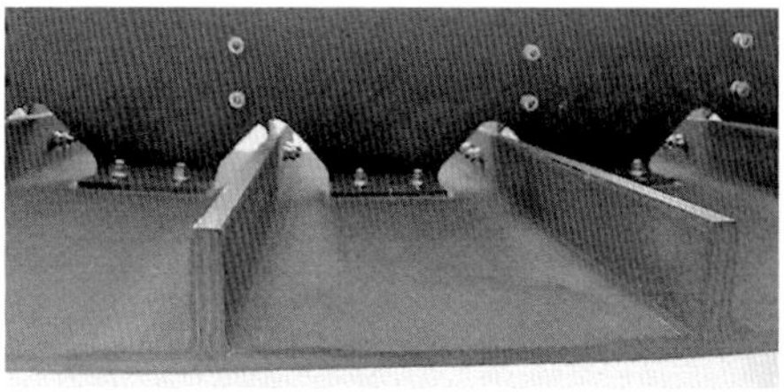

Figure 8.6
Skin panel with integrally bonded stringer profiles

Differential Design

Differential designs are multiple-step integration processes. Here, cured components, such as stringers, are combined with an uncured skin panel, and bonded to the skin panel while it cures (Figure 8.7). In a subsequent step, the ribs are integrated. This design concept is also called *rigid/soft* or *soft/rigid*, depending on which mating part is already cured.

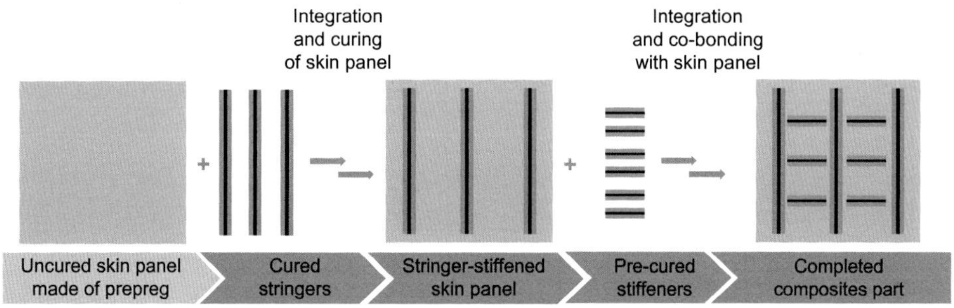

Figure 8.7 Multi-step integration process used for differential designs

The advantage of this design is the fact that all elements that are already cured prior to integration can be tested for defects. Therefore, these defective parts are rejected prior to integration, which is of immense importance for the large and very expensive components in the aircraft industry, where component costs reach US$ 40,000 to US$ 950,000. Another advantage are the considerably simpler molds for the manufacture of such panels. Here, individual components, such as stringer or skin panel, may undergo design changes without necessitating complete new tooling.

A disadvantage of this approach is the fact that the individual manufacture of stringers and ribs generally causes high additional costs. Component weight is also often higher than that achievable when using integral design concepts. In addition, after bonding to the prepreg skin, the pre-cured components have to be secured by rivets in order to meet aviation regulations, because the mating parts are adhesively bonded rather than cohesively.

Figure 8.8
Stringer on skin panel with additional rivets
[Courtesy: Airbus Operations GmbH]

8.1.3 Open Mold Concepts

Independent of the design concept (integral/differential concept), the curing of prepregs often takes place in open molds. The pressure required for the consolidation of the laminate is applied via an autoclave (see also Chapter 5). Both design concepts can also be used with closed molds, e.g., press-forming and SQRTM processes. We will describe the open mold variation in more detail, because it offers more flexibility in terms of integrating individual components (e.g., skin panel or stringers).

As detailed in Section 6.4.2, the open mold principle can be distinguished in three commonly used variations with respect to the different curing states of, e.g., stringer elements and skin panels:

a) rigid/soft (stringer cured, skin panel uncured)

b) soft/rigid (stringer uncured, skin panel cured)

c) soft/soft (both mating parts uncured)

Variations (a) rigid/soft and (b) soft/rigid are equivalent to the just-described differential design concepts; however, they are distinguished more accurately with respect to the state of curing of the mating parts. Variation c) soft/soft represents integral designs.

Depending on the curing state (a to c) of the stringer elements, additional tools are needed for their integration in order to meet the requirements for stringers on the skin panel in terms of shape- and position tolerances. The vacuum bag on the top surface of the component always creates closure towards the autoclave.

Rigid/Soft Variation (a)

The rigid/soft variation is suitable in particular for lightweight designs of stringer-reinforced skin panels, such as wing skins or vertical tail skins (Figure 8.9, left). Here, it is the goal to ensure that the stringers meet the required tolerances in terms of position and orientation on the skin panel. To this end, the stringers are placed with the help of a laser positioning device or an angled stop and then secured with adhesive tape.

The rigid/soft variation can be further distinguished in different set-up options.

Type 1: Vacuum creation via disposable film membrane

The vacuum film is draped over all stringer webs following the geometry so that it fits flush with all surfaces. Bridging of the bagging film between stringer web and skin panel must be avoided, because once autoclave pressure is applied (7 to 10 bar) it could lead to rupture of the film or a distortion of the stringer web (Figure 8.9, right). Either case would render the component defective. This variation is frequently used for large-area components because of its simplicity and the small number of necessary auxiliary materials. With correct drape of the disposable membrane (learning curve) it is possible to achieve very good results in terms of component quality and reproducibility.

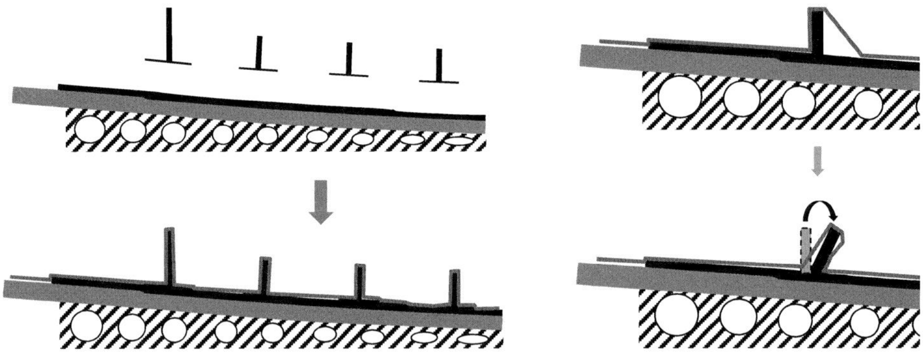

Figure 8.9 Vacuum creation with only vacuum bag (lefts), bridging of the bagging film (right)

Type 2: Vacuum creation via disposable film bag and silicone angle brackets

As a modification of Type 1, silicone angle brackets can be used at the stringer feet to facilitate draping of the disposable elastomer membrane (Figure 8.10). They prevent shifting or tilting of the stringer feet.

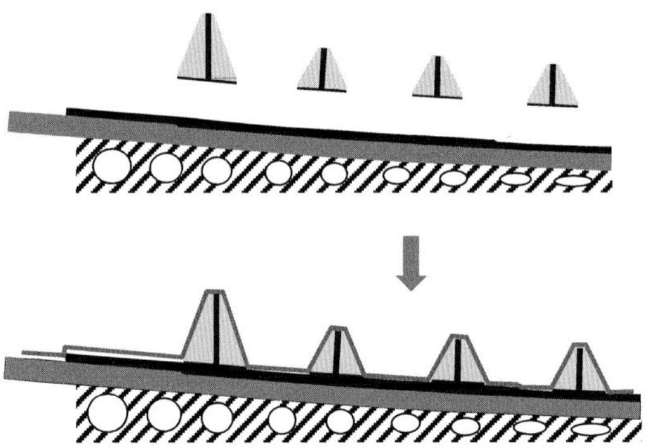

Figure 8.10
Vacuum creation via film bag and silicone angle brackets

Type 3: Vacuum creation via film bag and cut-out caul plate

Another open mold variation uses a caul plate made from metal or CFRP (Figure 8.11). Here, the caul plate facilitates positioning as well as securing of the pre-cured stringers on the prepreg stack during curing in the autoclave. This variation could almost be called a closed mold approach, because part of the top surface of the prepreg stack is completely covered by the shaping tool. However, analogous to all other variations, the seal is provided by a disposable membrane. Here, it would also be possible to employ a reusable vacuum bag made from silicone. The caul plate serves as a pressure sheet on the laminate. Depending on the type of caul plate (rigid or soft mandrel) it can cause the pre-cured stringer feet to press into the laminate. This typically leads to excess pressure in the prepreg stack below the stringer feet, which shifts the fiber/matrix ratio (typically 60:40) in favor of a higher fiber content. However, the caul plate creates a smooth, defined surface that may be required for subsequent assembly steps.

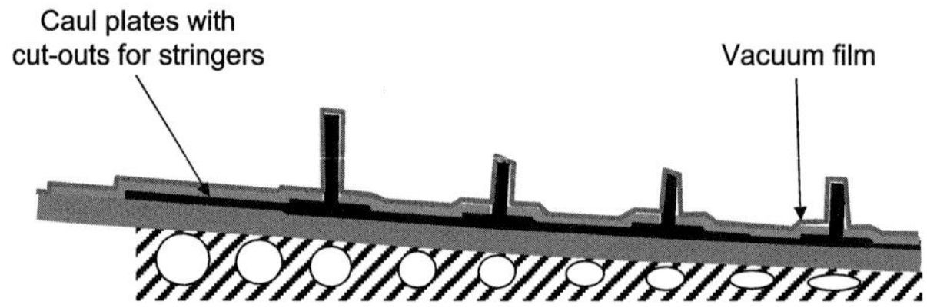

Figure 8.11 Vacuum creation via film bag and cut-out caul plate

Soft/Rigid Variation (b)

When soft/rigid variations are employed, uncured stringers (soft) are placed on a pre-cured prepreg laminate and subsequently cured (Figure 8.12). This method is particularly suitable for highly contoured components, such as underwing skins and horizontal stabilizers.

The T-shaped stringers are supported on both sides by brackets (segmented pressure intensifiers) made from either metal (Ni-36) or CFRP. These segments brace against the cured base laminate (Figure 8.13). Two matching brackets can be forced against each other with the help of tension clips. In order to meet position and orientation requirements, the bracket ends are often attached to the molding tool used for the base laminate. The seal (and vacuum creation) is provided by a disposable membrane, as with the other variations.

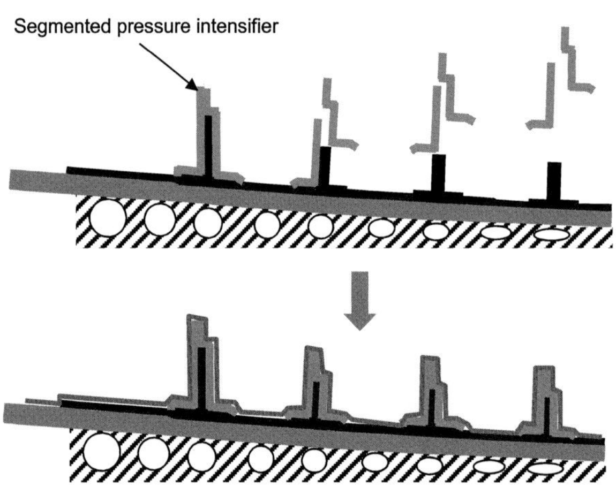

Figure 8.12 Vacuum creation with brackets made from metal or CFRP

Figure 8.13 Tooling for integration of stringers on CFRP-skin panel [Courtesy: Premium Aerotec GmbH]

Soft/Soft Variation (c) – Integral Design

The soft/soft variation is suitable for the bonding and forming/curing of two uncured component segments (e.g., stringer and base laminate), see Figure 8.14. It can be interpreted as a combination of caul plate and segmented pressure intensifiers (brackets). Here, curing mandrels (mainly Ni-36 or CFRP) are used to support the stringer feet and provide a defined and toleranced surface for the base laminate. The latter is often required to facilitate subsequent assembly steps. The same principle is applied for modular mandrels (Figure 8.15). In addition to rigid mandrels, rubber or silicone (soft mandrels) can also be used; however, because of their high wear, they are used less frequently.

In order to minimize their weight (in particular for Ni-36 mandrels) and to improve heating, curing mandrels may be fitted with bored chambers over their entire length (depending on diameter and length). As in the other variations, the seal is provided by a vacuum bag with disposable membrane. This variation is also called *integral design*, because various geometrical elements can be cured together in one step. Other modifications are also possible.

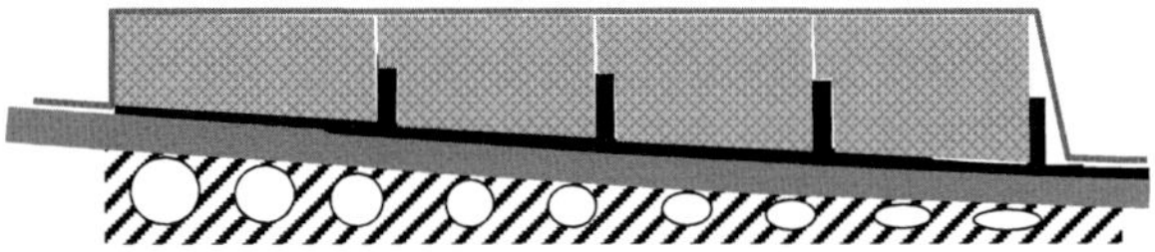

Figure 8.14 Vacuum creation with curing mandrel made from metal or CFRP

Often, additional curing mandrels are utilized; they can be either rigid or soft mandrels and are used to, e.g., position uncured stringers on an uncured skin panel within strict position and orientation tolerances (Figure 8.15) for subsequently curing.

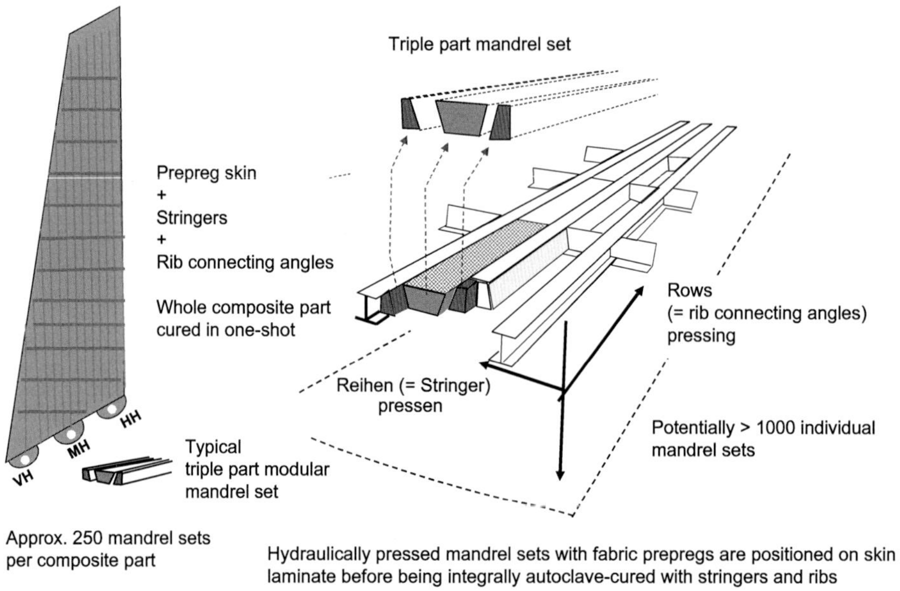

Figure 8.15 Curing mandrels for modular tail fins

Figure 6.18 shows a detail of such a modular tail fin. The double T-stringers were integrated in the skin panel in one step together with the ribs (rib connecting angle).

Figure 8.16 Detail of modular tail fin [Courtesy: Airbus Operations GmbH, Stade]

8.2 Determining Factors and Interactions

As mentioned in the introduction to this chapter, a variety of interactions between the parameters of the manufacturing process, such as material, construction, component design, tooling, and process, need to be considered. These parameters and their interactions will be evaluated and explained with the help of examples. Only optimal alignment of the individual parameters will result in process stability and zero-defect components. In the following we will use examples to evaluate and discuss some of the potential determining factors.

8.2.1 Determining Factors during Component Manufacturing

Figure 8.17 shows potential influences on the manufacturing process and the final component. It also shows clearly that the major influencing factors, such as the prepreg material, include a number of other parameters.

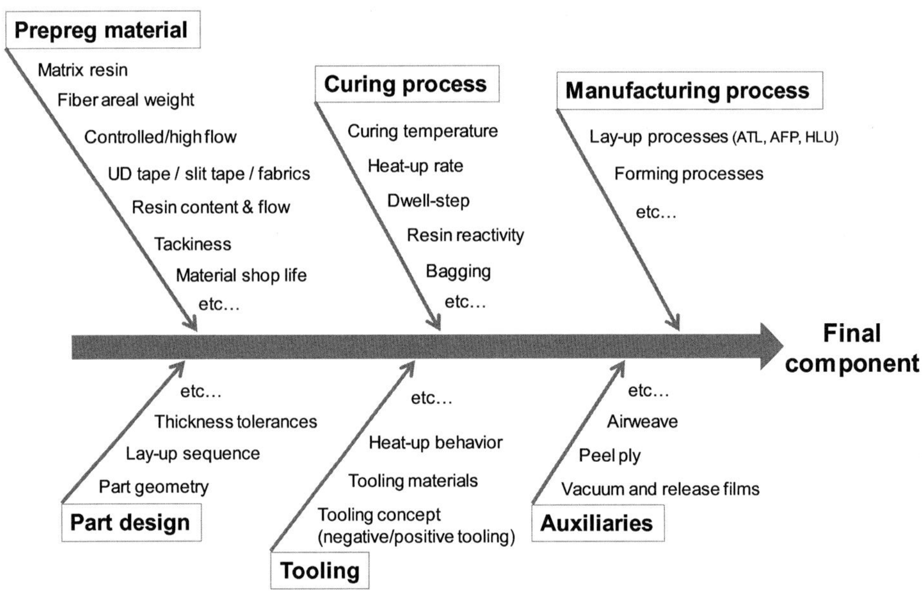

Figure 8.17 Parameters influencing the manufacture of components using prepregs

It is difficult to evaluate the individual factors influencing the manufacturing process because of the interrelations between the individual parameters. A simplified diagram of these interrelations is depicted in Figure 8.18. With the selection of a prepreg material for a component design and its structural configuration, the manufacturing process e.g., ATL or AFP) as well as the curing process are also already defined.

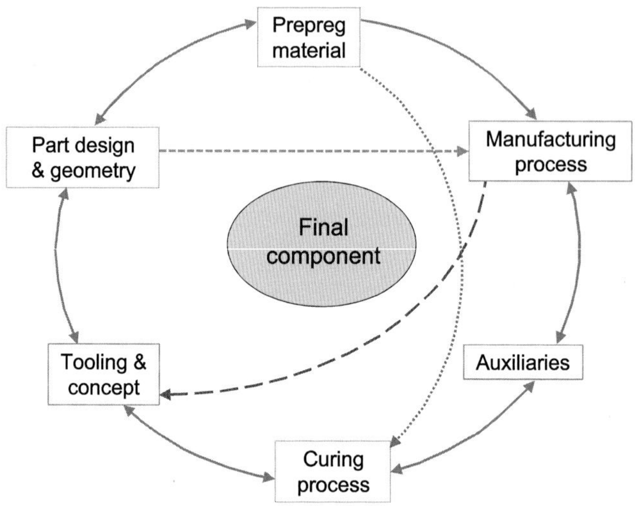

Figure 8.18 Potential interrelations of parameters

In turn, both manufacturing process and component design affect the tooling concept. The component design already establishes which regions of the component require a defined surface and laminate thickness in order to facilitate the subsequent assembly steps. Component size and the selected manufacturing technology determine the design concept (see above) and the tooling. Here, specific attention should be paid to the productivity of the entire manufacturing process. The component's curing process is often an underestimated bottleneck. For example, the utilization of metal tooling causes slow heating and cooling rates and thus long curing cycles (see also Section 5.2). The selection of auxiliary materials may even aggravate this effect because of potential additional thermal insulation. At the same time, the curing cycle (slow or fast heating) together with the utilized auxiliaries (peel ply, air weaves) affect the internal laminate quality. Adverse interactions between process control, tooling, and materials may cause problems such as porosity, waviness, fiber warpage or even cracks in the component. Also encountered are too much or too little resin flow with the associated deviations from specifications in laminate thickness.

The experiences gained over approx. 70 years of composite manufacturing allow relatively good assumptions. However, the determining factors do not always interrelate as expected or desired.

8.2.2 Interrelations during Component Manufacturing

In the following, we will use three examples to evaluate the correlations between determining factors and their potential causes and consequences for the manufacture of CFRP components. These examples are based on practical experiences, not scientifically substantiated findings. Each manufacturing process for CFRP components exhibits individual, seemingly small, differences. The complex relationships are often not obvious so that the described causes may not be the only initiators but rather should be considered as explanations of the phenomena encountered.

Interrelations during Manufacturing of U-Profiles

The mutual effects of component design, material, tooling concept, and curing cycle will be described using the manufacture of U-profiles as an example (Figure 8.19). Two U-profiles may be used, e.g., to form a double T-stringer.

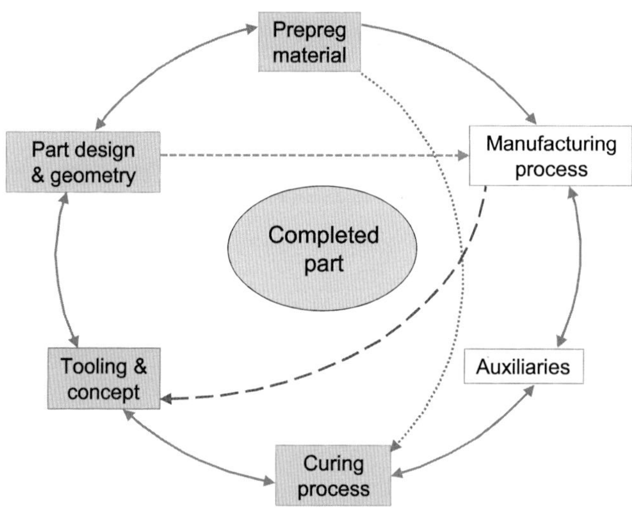

Figure 8.19 Effects of design, material, tooling, and curing process on component

These stringers are often used as reinforcing elements for skin panels in the aircraft industry. Here, they have to provide high stiffness in the axial direction (in the web), while at the same time they should be lightweight. The stiffness is achieved, e.g., through geometrical parameters such as web height and laminate thickness in stringer web and foot. In order to meet these requirements, the laminate structure for the U-profile has a high content of 0°-layers in the axial direction, i.e., in the web. Increasing the 0°-layer content in the stringer web is a deliberate deviation from a quasi-isotropic laminate structure (Figure 8.20). If a quasi-isotropic laminate structure were chosen (i.e., same content of 0°, 90°, and

±45°-layers, Figure 8.21), the same stiffness would require higher laminate thickness and thus lead to higher weight.

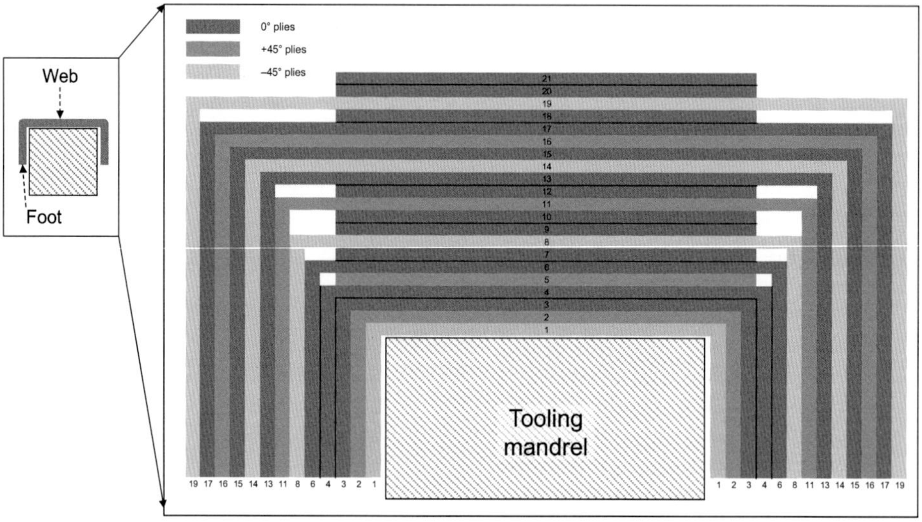

Figure 8.20 Layer build up for U-profile with high 0° content in web

Based on a prepreg areal weight of 194 g/m², removal of the additional 0°-layers in the web will reduce the weight in the feet by approx. 1.5 kg/m².

Hot forming (Chapter 4) is used to mold stacks of flat prepregs over a mandrel, made, e.g., from Ni-36, into U-profiles. During the curing cycle, the mandrel remains in the component to facilitate molding. For the manufacture of double T-stringers, two U-profiles are bonded back to back in an autoclave while curing. The created double T-profile is subsequently separated along the web in two equal halves.

The choice of material for the curing mandrels, in combination with the already determined layer structure, affects the interrelation of component and tooling during the curing cycle. As long as the prepreg stack is uncured, it is able to move freely with the expansion of the curing mandrel during the heating phase. Once the laminate is cured, it exhibits its own thermal expansion characteristics, which may differ sharply from the expansion characteristics of the metal tooling, depending on layer structure. Figure 8.21 shows the thermal expansion behavior of various metals as well as that of CFRP laminates as a function of the selected ply orientation and position.

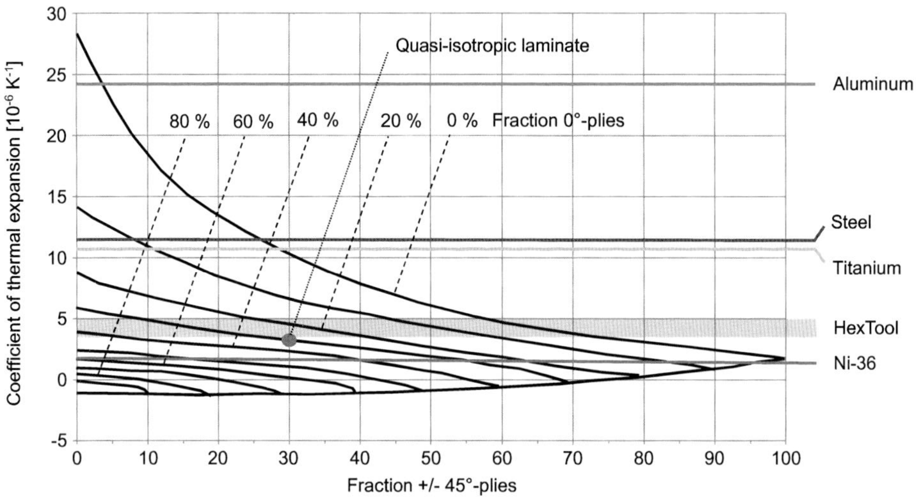

Figure 8.21 Coefficient of thermal expansion of CFRP as a function of ply orientation and position

The nomogram in Figure 8.21 shows the expansion behavior of CFRP laminates as a function of layer content in ± 45° and 0° orientation. The 90°-layer content can be calculated based on the other layer contents. The red dot marks a quasi-isotropic (QI-)laminate structure that consists of approx. 30% layer content in each orientation. The *y*-axis shows the coefficient of thermal expansion (CTE) for each laminate structure. The figure shows clearly that the CTE of a QI-structure is close to the CTE of Ni-36 and HexTOOL®. However, for the layer structure described in the example above, the CTE in axial direction is negative, while in lateral direction (transverse to the U-profile axis) the CTE is higher than those of HexTOOL® or Ni-36 (Figure 8.22).

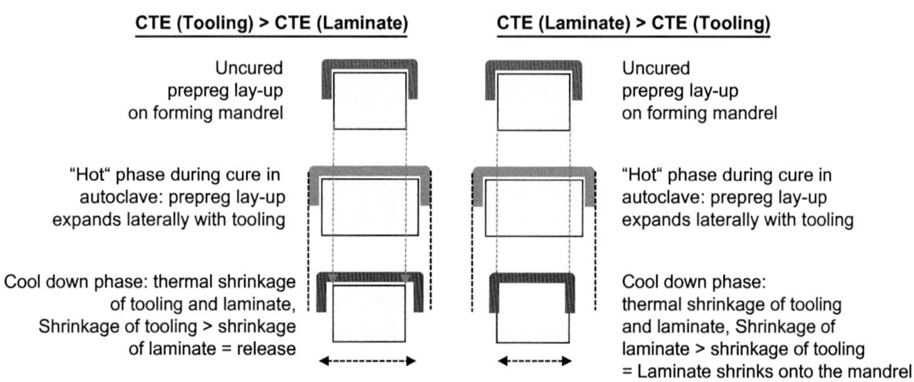

Figure 8.22 Thermal shrinkage of tooling and laminate

The latter means that the cured laminate will shrink itself onto the selected tooling material during curing. This may cause significant stresses and lead to cracks in the fiber layers (Figure 8.23). The mismatch in shrinkage behavior can be abated by the selection of a tooling material with a coefficient of thermal expansion higher than that of the component, which will also simplify the demolding operation of mandrel and U-profile.

Another approach to mitigate this effect is a change in prepreg stack design. Instead of a U-shaped stack, two L-shaped stacks could be processed on one mandrel. The separation avoids stresses caused by thermal expansion and shrinkage. However, the process has to be re-evaluated in terms of automation, potentially reduced flexibility, longer cycle times, as well as achievable accuracy of stack positioning on the mandrel.

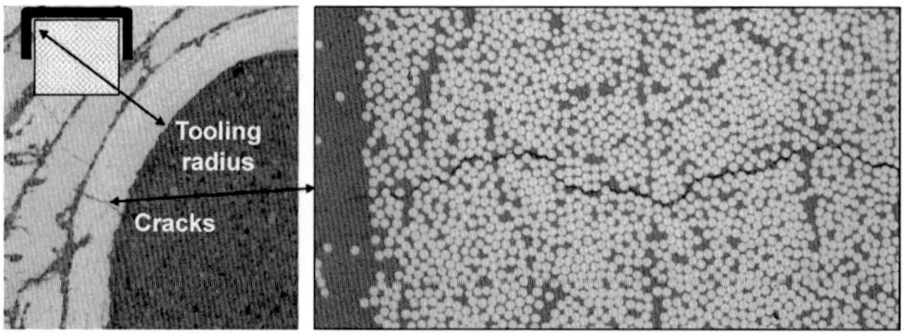

Figure 8.23 Interfiber fracture in radius area

The tooling design may also create additional challenges. To make sure that the tooling halves are properly aligned, a metal pin can be attached to the end of the upper (or lower) mandrel that will fit into the respective bushing in the other mandrel half (Figure 8.24).

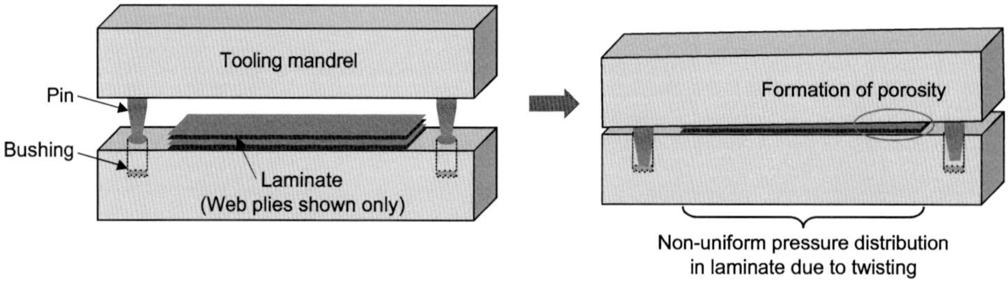

Figure 8.24 Misalignment of curing mandrels and porosity formation

If the tolerances for pin and bushing (fitting sleeve) are set too narrow, the mandrels may already be twisted at the initial contact so that the upper mandrel can no longer freely move in *z*-direction. This effect may even intensify with thermal expansion during the curing cycle. The limited mobility and/or twisting may cause slight differences in pressure distribution in the laminate placed between the mandrels and thus to the formation of porosities in certain regions of the component. This effect can be mitigated by utilizing fixed/floating bearing arrangements. The fixed bearing can be attached on one end of the mandrel in the form of a pin/bushing combination, while the floating bearing can be located at the other end of the mandrel in the form of an elongated hole/pin combination that creates a flexible bearing and prevents twisting. The pins may also take spherical forms on both sides of the mandrel, providing a geometry that does not facilitate twisting, in contrast to straight pins.

Interrelations: Example of a Structure Manufactured Using a Negative Design

The manufacturing process of a negative-design component (Figure 8.25, right) will be used in the following to describe the mutual influence of component geometry, material, and tooling concept.

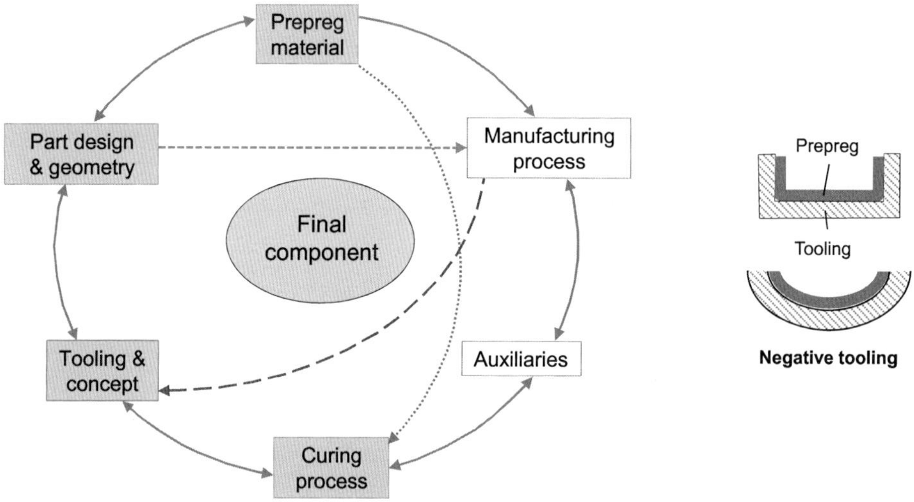

Figure 8.25 Effects of geometry, material, and tooling concept on component

Here, the design specifications require a smooth outer surface of the component (toleranced surface) for subsequent assembly; therefore, a negative tooling design is selected. The prepreg material is placed in the mold cavity and the sides are shaped to create a U-profile. In order to increase tack for better processibility (Chapter 3), a lower level of impregnation may be selected, i. e., a larger proportion of the prepreg resin is located on the surface of the fiber bed rather than wetting the whole fiber layer (Figure 8.26). This partial impregnation results in a larger "wet thickness" of the individual prepreg ply than the cured ply thickness (CPT).

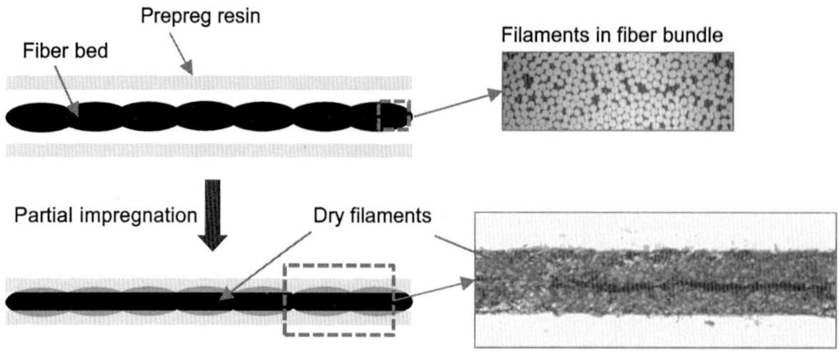

Figure 8.26 Prepreg with low level of impregnation

When several such prepreg plies (e.g., up to 20 to 30 individual plies) are laminated, the "wet thickness" effect adds up markedly so that during the heating phase in the autoclave, the prepreg stack undergoes consolidation in its thickness direction (Figure 8.27) and thus an in situ post-impregnation (Figure 8.28).

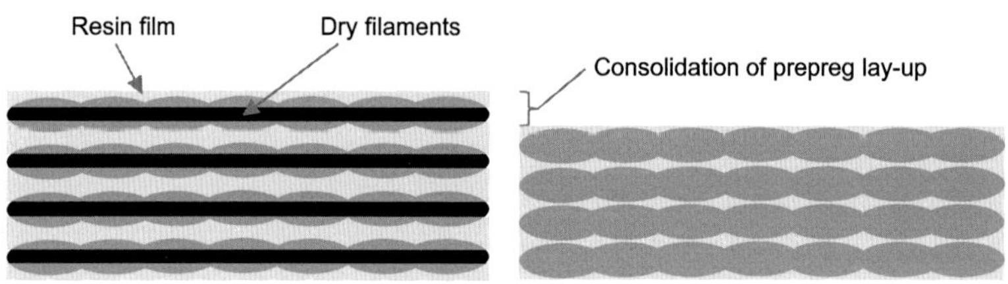

Figure 8.27 Consolidation of prepreg stack

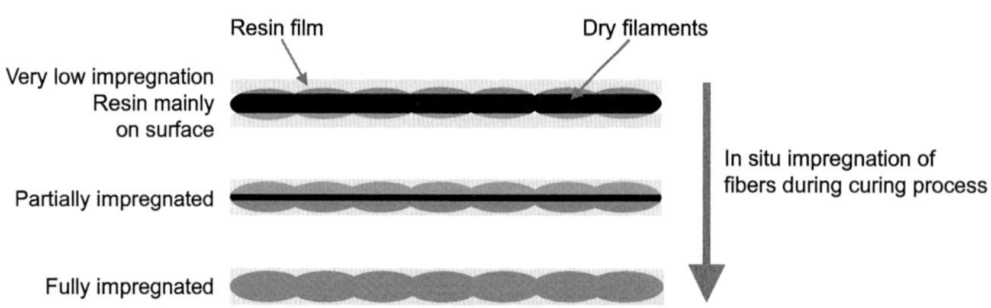

Figure 8.28 In situ impregnation during curing cycle

This consolidation in thickness direction poses a problem because the laminate is compacted in all directions simultaneously. Here, the plies further away from the tooling (inner plies) travel a larger distance when they are pressed against the tooling and the left and right. At the same time, there is also consolidation in the downward direction (Figure 8.29).

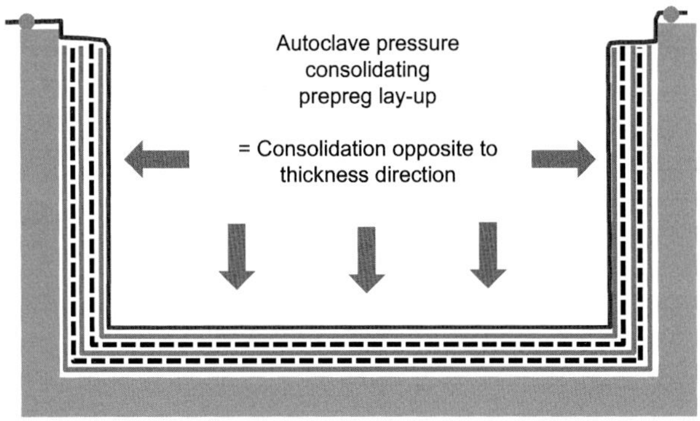

Figure 8.29 Consolidation and compaction inside tooling

Thus the inner plies are forced to expand or to perform a relative movement in order to balance the change in length caused by the consolidation. However, they may not always be able to make these adjustments. The applied autoclave pressure (typically 7 to 10 bar) and the bonding between the plies by the prepreg resin render relative movement of individual plies impossible. Because the autoclave pressure is more effective on the plane than on the radii, the inner prepreg plies will bridge the radius to a certain degree, thus changing force progression and load capacity of the component (Figure 8.30). The continuous, inner line marks the actual, the dotted line the required fiber orientation.

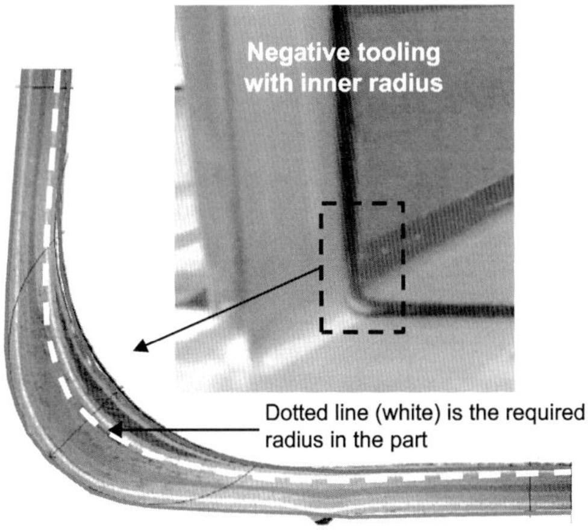

Figure 8.30
Micrograph of radius area showing the bridging effect of inner plies

This problem can be mitigated by one of several measures. For one, using a prepreg with a higher level of impregnation may minimize the consolidation. Another approach is the manufacture of a flat prepreg stack via ATL (rather than hand lay-up) and subsequent hot forming that would facilitate a pre-compaction of the stack prior to the curing process. Yet another, although somehow controversial, method is the use of pressure strips (rubber or silicone corner profiles) in the radii. It is possible to effectively increase the autoclave pressure in this area using pressure intensifiers in the form of round cords. However, they often cause deep undesired indentations in the laminate or the formation of beads in the border area of the pressure strips.

Interrelations: Example of a Sandwich Structure

The interrelations between component design, material, and curing process and their implications will be described using a sandwich structure as an example.

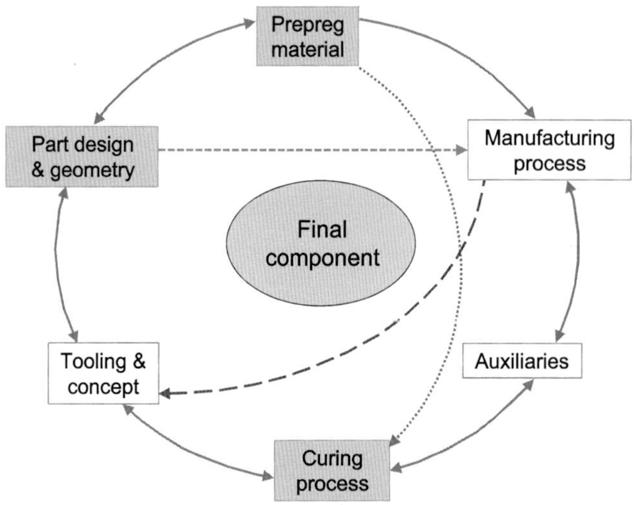

Figure 8.31 Effects of design, material, and curing process on component

PMI rigid foams (Evonik) can be used for the manufacture of Omega stringers (Figure 8.32). Typically, the formed foam remains in the component after curing of the prepreg, allowing for one-step curing of stringer and skin segment to be reinforced.

The cell walls of certain rigid foams do not exhibit sufficient dimensional stability under the temperatures (typically 180 °C) and pressures (7–10 bar) required for prepreg curing. Therefore, the pressure on the component is reduced to, e.g., 3 bar absolute.

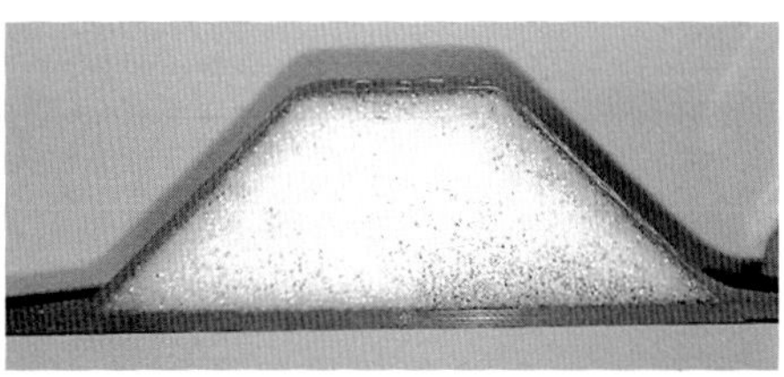

Figure 8.32 Example of sandwich design: Omega-stringer with foam core

However, depending on the prepreg matrix high pressure (7–10 bar) is required during the heating and curing cycle in order to initiate shear forces on the prepreg and thus achieve sufficient resin flow. The resin flow makes sure that air trapped between prepreg plies or between prepreg and foam core is flushed out with excessive resin, thus facilitating void-free components.

The reduced pressure in the sandwich component may lead to reduced resin flow and in turn to void formation, in particular between unidirectional plies (Figure 8.33). Void formation is further promoted by slow or fast heating in the autoclave as well as by unsuitable vacuum creation.

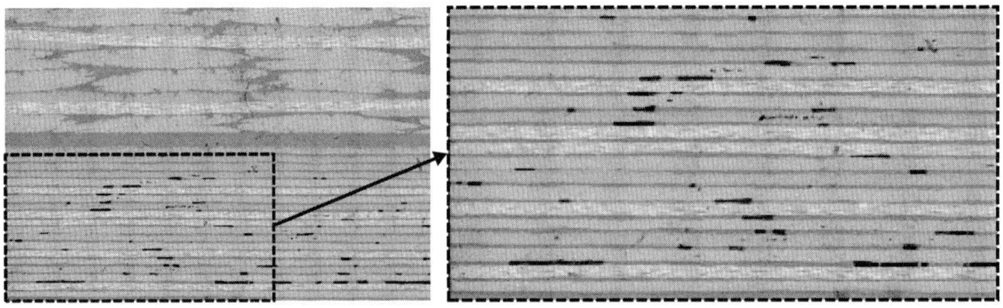

Figure 8.33 Voids in the UD-layers of a sandwich component

The problem can be solved by a specially adapted autoclave cycle (heating ramp, pressure) and the utilization of pressure-resistant PMI foams. Another option is the utilization of metal mandrels for the forming operations.

Stringer-Reinforced Skin Segments

A good example for the optimal interaction between the individual parameters are stringer-reinforced skin panels, manufactured using either integral or differential design open mold concepts (Section 8.1). For more than 20 years, these concepts have been successfully used in the aircraft industry to manufacture large components, such as skins for vertical tails and, more recently, also wing fairings. With growing experience in the manufacturing of composite components, material, design, and manufacturing, including curing, have been adjusted and coordinated. In the early stages of development and manufacturing of these components, the lack of experience and of automated processes left manual processes as the technology of choice. Here, the "human factor" caused large variations in the final components. However, the knowledge gained during this period benefits the development of composite components today. In particular appropriate component design tailored to fiber reinforced materials has made great strides, considering the fact that in the beginning these components were designed using the principles traditionally used for metal components, thus not taking advantage of the potential CFRP materials are offering. With respect to processibility limitations of prepregs materials, broad knowledge and versatile skills were gained over time which today allow for series production of, e.g., aircraft components. Current examples are the latest aircraft generation with fuselages made form CFRP for Airbus (A350XWB-900) and Boeing (B787-8).

Figure 8.34
CFRP fuselage of Airbus A350XWB [Courtesy: Airbus Operations GmbH]

8.3 References

[1] G. Niederstadt. *Ökonomischer und ökologischer Leichtbau mit faserverstärkten Polymeren.* Expert-Verlag, 1997.

[2] U. Bieling. *Serieneinsatz von Faserverbundwerkstoffen im Flugzeugbau, dargestellt am Seitenleitwerk des Airbus* (Talk). Prague: 10. ICED-Conference, 1995.

[3] H. Apmann. *Vorteile durch CFK und Leichtbau.* Company presentation, Premium Aerotec GmbH, 2013.

[4] MIL-Handbook-17-1E. *Materials and Processes – The Effects of Variability on Composite Properties.* 2002, Chapter 2, pp. 1–69.

9 Conclusion

In the future, a more integrated approach has to be found for the entire value chain representing the manufacture of fiber reinforced components to make high-performance fiber reinforced plastics more competitive with metals and to introduce them to wider markets. This approach has to include all early process stages, such as design of matrix, textiles, and component geometry, as well as automation strategies for processes that are currently based on manual lay-up, and assembly operations. This means that material performance, process requirements, manufacturing, and functionalities of the component have to be seen holistically. Another focus area is the development of eco-efficient recycling strategies in order to minimize unnecessary use of expensive raw materials and/or creation of scrap. It is expected that research and industry will decidedly advance these issues in the next 5 to 10 years.

There are already first successful approaches for the utilization of recycled carbon fibers in new applications and components. Figure 9.1 shows a sandwich component prototype in which the face sheets are made of carbon fiber mats from recycled carbon fibers. Possible applications are cover panels in the automotive and aircraft industries, such as lids for overhead bins in aircrafts.

Figure 9.1 Recycled fibers in a new application as face sheet of a sandwich component

The Authors

Dr.-Ing. Hauke Lengsfeld

After completing his Masters in Chemistry and his Ph.D. at the Department of Polymeric Materials at the University of Bayreuth (Germany) (Adviser: Prof. Dr.-Ing. Altstädt), Dr.-Ing. Hauke Lengsfeld was responsible for the production of large CFRP components at Airbus Operations GmbH as a materials and process expert. He was project manager for the development of CFRP components for the Airbus A350XWB. Subsequently, he was Head of Technical Marketing and Sales at Saertex Stade GmbH. Since 2010 he has been the Technical Support Manager for Hexcel Composites GmbH in Central Europe. Since 2007, Dr.-Ing. Hauke Lengsfeld is a adjunct professor, teaching courses on fiber reinforced materials at several universities.

Prof. Dr.-Ing. Volker Altstädt

Volker Altstädt undertook university studies in Physics and received a Dr.-Ing. in 1987 with Prof. Dr.-Ing. Ehrenstein at the department of Mechanical Engineering at the University of Kassel (Germany). For eight years he was head of the Department of Polymeric Physics in the Division for Polymer Development at BASF AG in Ludwigshafen. Since 1995 he has been professor for Plastics in Mechanical Engineering and head of the Department of Polymers and Plastics Composites at the Technical University of Hamburg-Harburg. Since October 2000 he has been professor in the Department of Plastics Engineering at the University of Bayreuth.

Dr.-Ing. Felipe Wolff-Fabris

Dr.-Ing. Felipe Wolff-Fabris studied material science at the Universidade Federal do Rio Grande do Sul in Porto Alegre (Brazil) and at the Ecole Centrale de Nantes (France). After graduation, Dr.-Ing. Wolff-Fabris worked at the Department of Polymeric Materials at the University of Bayreuth leading the research group on "Thermosets and Fiber Reinforced Materials". He worked as laboratory manager for "Composites" at Henkel AG and subsequently as lead engineer at the University of Bayreuth. Since 2012, Dr.-Ing. Wolff-Fabris has headed the European Center for Dispersion Technology at the Kunststoff-Zentrum SKZ.

Dipl.-Phys. Johannes Krämer

Johannes Krämer earned his Masters degree in physics and worked at the Department of Polymer Engineering at the University of Bayreuth under Prof. Dr.-Ing. Altstädt in the field of thermosets and fiber reinforced plastics. Since 2012 he has worked for BASF SE in the Department of Structural Materials, advanced materials and systems research in Ludwigshafen.

Dr.-Ing. Javier Lacalle

After attaining his Master in mechanical engineering at the Universidad Pública de Navarra in Pamplona (Spain), Dr.-Ing. Javier Lacalle received his Ph. D. in 2005 from the Department of Mechanics at the University of Duisburg (Germany) (Advisor: Prof. Dr. rer. nat. Manfred Braun). Until 2011 he worked at M. Torres Diseños Industriales as project and product manager for lay-up technologies. In 2010 he received an additional Master of Science in the area of composite materials from the PFH-University Göttingen in Stade. Between 2011 and 2014 he was group manager for automated CFRP processes and for AFP technology at the Fraunhofer Institute for Chemical Technology (ICT-FIL) in Augsburg and in the Department for Carbon Composites at the Technical University in Munich. Since May 2014 he has been project manager and technical consultant at M. Torres Deutschland.

Index

Symbol

A

B

C

V

W

Z